Building Research Establishment

Aspects of fire precautions in buildings

R E H Read, FRICS, MIFS
W A Morris, BA, MIFS

Fire Research Station
Building Research Establishment
Borehamwood, Herts
WD6 2BL

Prices for all available
BRE publications can be
obtained from:
BRE Bookshop
Building Research Establishment
Garston, Watford, WD2 7JR
Telephone: 0923 664444

BR 225
ISBN 0 85125 533 7

© Crown copyright 1983, 1988, 1993
First published 1983
Second edition 1988
Third edition 1993

Applications to reproduce extracts
from the text of this publication
should be made to the Publications Manager at
the Building Research Establishment, Garston

Contents

	Page
List of figures	vii
List of tables	viii
Acknowledgements	ix
Foreword	x
1 Introduction	1
2 Historical background	8
Legislation	8
— Middle Ages	8
— Seventeenth Century	9
— Eighteenth Century	11
— Nineteenth Century	12
— Twentieth Century	17
Fire testing	21
Bibliography	27
3 Existing legislation and other requirements	28
Legislation	28
— Building Regulations	28
— Requirements in connection with occupied premises	31
Fire insurance requirements	40
4 Understanding fire	41
The growth of fire	42
— Ignition	42
— Fire growth	45
The developed fire	47
The relationship between the fire and fire tests	47
— Tests which measure ease of ignition	48
— Tests concerned with the growth of the fire	48
— Tests involving the developed fire	49

(continued)

5	**Structural fire protection**	51
	Standard fire tests	51
	Fire growth	53
	Fire spread	57
	— Internal	57
	— External	57
	Structural stability	60
	Fire resistance of elements of building construction	60
	— Concrete, steel and timber structures	62
	— Glazed elements	63
	— Walls	64
	— Floors	64
	— Roofs	65
	Fire doors	65
	Smoke movement	69
	Roof venting	71
6	**Means of escape in case of fire**	74
	Requirements	75
	Considerations	75
	— Type of risk and number of occupants	76
	— Number, location and size of escape routes	76
	— Travel distances	79
	— Protection against spread of smoke and heat	79
	— Doors across escape routes	80
	— Signposting and lighting of escape routes	80
References		82
Appendix 1	**Extracts from historical documents**	91
	A. London Building Act 1894 (The Second Schedule)	91
	B. BFPC Standards of fire resistance	92
Appendix 2	**Extracts from current legislation**	94
	A. London Building Acts (Amendment) Act 1939	94
	B. Fire Precautions Act 1971	95
	C. Fire Certificates (Special Premises) Regulations 1976	99
	D. Building Act 1984	101
	E. The Building Regulations 1991	103
Appendix 3	**Standard fire tests**	105
	BS 476: Fire tests on building materials and structures	105
	BS 2782: Methods of testing plastics	118
Appendix 4	**Reading list and other relevant sources of information**	121
Index		130

List of figures

Frontispiece Aftermath of Bradford City football club fire

Figure		Page
1	Estimated fire claims for England, Scotland and Wales, 1978–1992	2
2	Sections of houses authorised by the Rebuilding Act of 1667	10
3	Fire escapes with fly ladders and chutes in action (circa 1850)	14
4	Tooley Street fire 1861	15
5	Hartley's house at Putney Heath	21
6	General view of the BFPC's testing station near Westbourne Park	23
7	A group of the members of the 1903 International Fire Prevention Congress	25
8	Greater London	30
9	Fire — a simple model	42
10	Heat transfer and heat loss mechanism	44
11	Influence of substrate	46
12	Initial fire growth	46
13	Fire phenomena and BS 476 Part numbers related to the stages of an uncontrolled fire in the compartment of origin	52
14	Effect of wall lining on time to flashover in test rooms	53
15	Radiation	58
16	External fire exposure	59
17	Underside of timber joisted floor construction after test, with ceiling protection removed	66
18	'Any attempt to prevent a fire door closing ...'	68
19	The forces driving smoke through a building	70
20	Roof venting	71
21	Screens beneath a flat roof	71
22	Effect of venting on flames beneath a ceiling	73
23	Examples of fire safety signs	81
24	External fire exposure roof test (BS 476: Part 3: 1958)	105
25	Relation between the roof test and a fire	106
26	Non-combustibility test apparatus (BS 476: Part 4)	108
27	Ignitability test (BS 476: Part 5)	109
28	Fire propagation test apparatus (BS 476: Part 6)	109
29	Surface spread of flame test (BS 476: Part 7)	110
30	An indication of classification limits for the spread of flame test	111

(continued)

31	Ignitability test (BS 476: Part 13)	112
32	Standard temperature/time curve (BS 476: Part 20)	113
33	Lighting wall furnace prior to test of loadbearing specimen	115
34	Column furnace showing a steel column after test	115
35	Floor furnace showing a timber floor collapsing	116
36	BS 476 Section 31.1 — diagrammatic arrangement of apparatus	117
37	BS 2782 Method 120A — diagrammatic arrangement of apparatus	118
38	BS 2782 Method 508A — diagrammatic arrangement of specimen and apparatus	119
39	BS 2782 Method 140D — diagrammatic arrangement of apparatus	120
40	BS 2782 Method 140E — general arrangement of apparatus	120

List of tables

Table		Page
1	Fires by location, 1981–1991	5
2	Fires in dwellings and other occupied buildings by cause, 1981–1991	6
3	Fires in occupied buildings by spread of fire, 1981–1991	7
4	Casualties from fires by location group, 1981–1991	7
5	Statutory provisions relating to occupied premises	32
6	Ignition temperatures	43
7	Fire propagation indices for typical materials	55
8	Some typical results of surface spread of flame tests	55
9	Examples of different 'half-hour' periods of fire resistance	61
10	Provisions as to method of test and minimum period of fire resistance for walls (Approved Document B, 1992 Edition)	61
11	Charring rates of timber	63
12	Capacity of a stair for total evacuation of a building	78
13	Examples of maximum travel distances permitted	78
14	Classification on roof test (1958)	107

Acknowledgements

The authors wish to express their gratitude for the many helpful comments given by their friends and colleagues — past and present — at the Fire Research Station, British Standards Institution, Department of the Environment, Department of the Environment (Northern Ireland), Fire Protection Association, Health and Safety Executive, Home Office, London District Surveyors Association, Loss Prevention Council Technical Centre and the Scottish Office.

Acknowledgement is also made to the following:

Fire Protection Association	— Figures 3, 4 and 15
Firehouse Magazine, 515 Madison Avenue, NYC, NY	— Figure 5
BT Batsford	— Figure 6
British Standards Institution	— Figures 13, 23, 30 and 32
British Standards Institution	— Tables 11 and 12

Foreword

Fire precautions are the measures taken and the fire protection provided in a building or other fire risk to minimise the risk to the occupants, contents and structure from an outbreak of fire.

This book is intended to provide an introduction to students and others interested in the passive aspects of fire precautions. It looks at the development of legislative requirements and tests since the Middle Ages, the nature of fire, structural fire protection techniques and means of escape in case of fire. Measures designed for detecting or for extinguishing fires and the provision of facilities for use by the fire service are not covered.

In this revised edition, the chapter dealing with 'Existing legislation and other requirements' reflects the recent changes to the building regulations and to the control of fire precautions in occupied premises. Opportunity has also been taken to update the other chapters where appropriate.

Current Statutory Instruments, British Standards and other relevant guidance documents are listed in Appendix 4. However, it should be stressed that as revisions are frequently being made and additional guidance published, such an Appendix cannot be expected to remain up to date.

R E H Read and W A Morris
Fire Research Station
13 July 1993

Chapter 1: Introduction

Over the centuries fire has continually caused injury, loss of life and destruction of property — particularly in urban areas. The more notable fires such as the Great Fire of London (1666) have destroyed large areas; or caused many deaths in one building, such as the fires at Summerland in the Isle of Man (1973), the Manchester Woolworth's store (1979), the Stardust Disco in Dublin (1981), Bradford City football ground (1985) and King's Cross (1987).

In the United Kingdom there are over 14,000 casualties (including over 800 fatal) a year in fires — mainly in buildings; with the majority of casualties occurring in dwellings. Each year the Home Office publishes annual fire statistics compiled from reports submitted by local authority fire brigades. Tables are provided based on the date, location, source of ignition, material or item first ignited, room or place of origin, spread, and method of extinction of fires, etc. A commentary on the statistics and a review of the last decade or so is also given. Tables 1–4 provide examples of the information contained in the published UK fire statistics for 1991[1].

Every quarter the Association of British Insurers (ABI) releases its figures for estimated fire claims incurred for material damage in Great Britain. In 1992 these cost over £800 million. Each month the 'Fire Prevention' journal gives an analysis of fires which caused losses of £250,000 or more (a yearly analysis is also made of £50,000-plus fires). Figure 1 shows the annual totals for fire claims estimated for England, Scotland and Wales for 1978–1992.

Direct damage by fire can occur not only in buildings but to equipment, work-in-progress, stores, records and other contents. Damage also arises from flame, heat, smoke and the water used in firefighting. Consequential or indirect losses such as interruption to business, costs of temporary reorganisation, lost orders, etc may often be far higher than the direct fire damage. Many firms go out of business following quite small fires.

The Fire Grading Committee in 1946 explained in Part I[2] of its report that in order to develop a rational system of fire precautions it is first necessary to analyse the various risks or 'hazards' to which fire may give rise. These hazards can be subdivided into:

Internal: those hazards which arise inside the building and which concern the occupants (ie Personal Hazard), and the structure and contents (ie Damage Hazard).

External: those hazards which arise as a result of fires in surrounding property (ie Exposure Hazard).

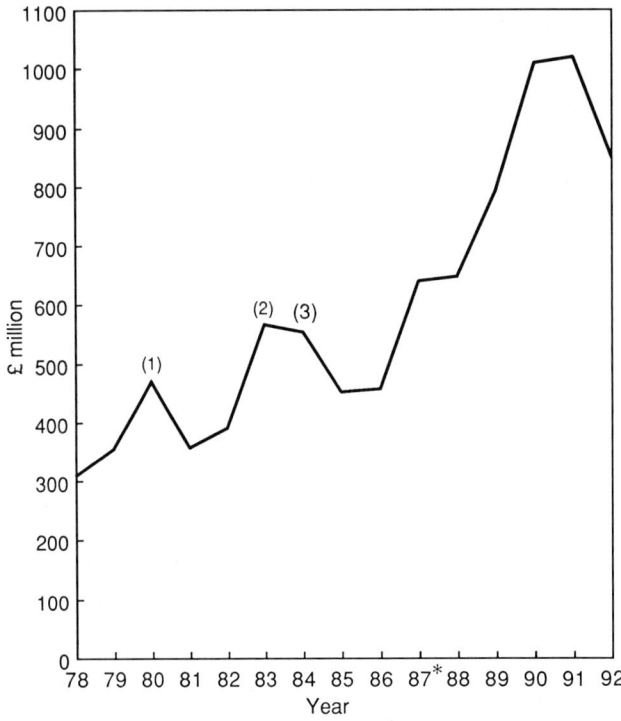

*The ABI calculation method was modified in 1987

(1) Including British Aerospace, Weybridge (£72.5 million) and Alexandra Palace (£31 million)
(2) Including Army Ordnance Depot, Donnington (£165 million)
(3) Including two London warehouses totalling over £81 million

Figure 1 Estimated fire claims for England, Scotland and Wales 1978–1992

The relative importance of each of these hazards will therefore vary according to the purpose for which the building is used, eg whether the building concerned has a large number of occupants such as a place of assembly or whether it is used mainly for storage.

The Committee therefore stated that the object of fire precautions is to minimise fire hazard and this could be achieved by:

(i) reducing the number of outbreaks of fire;

(ii) providing adequate facilities for the escape of the occupants, should an outbreak of fire occur;

(iii) minimising the spread of fire both within the building and to nearby buildings.

'Fire precautions' can be divided into *fire prevention* and *fire protection*. These are defined in BS 4422: Part 1[3] as:

fire prevention, 'measures to prevent the outbreak of a fire and/or to limit its effects',

fire protection, 'design features, systems, equipment, buildings, or other structures to reduce danger to persons and property by detecting, extinguishing or containing fires'.

Hence 'fire protection' accepts the possibility that the efforts directed towards 'fire prevention' will at some time fail. *Passive* fire precautions are those aspects of design which are intended to ensure the integrity of the building and the safety (including evacuation) of the occupants in the event of fire; whereas *active* fire precautions are those measures which operate only after ignition has occurred, such as detection and extinction.

Before turning to the following chapters the importance of the correct understanding of technical expressions should be emphasised — particularly in those fields where legislative requirements are concerned. The subject of fire is one in which it is usual for people such as the building control officer, fire prevention officer, architect, surveyor, builder, insurers and client to be directly involved in making decisions, or in meeting specifications, and yet these people will frequently have different interpretations of terms used and therefore different opinions as to what exactly is required.

In 1969 the British Standards Institution (BSI) published the first part of BS 4422 which consisted of 'terms and definitions intended for general application in the field of fire engineering, fire prevention and fire technology'. This glossary, which is currently being revised following the production by the International Organization for Standardization (ISO) of a vocabulary of fire protection (ISO 8421, see Appendix 4), provides a set of simple basic definitions which will be correct in any context but each of which is capable of extension for specific purposes for use by the specialist. Terms having normal dictionary meanings which are acceptable and sufficiently specific in the context of fire are not included. However, it should be noticed that occasionally ordinary expressions have distinct meanings in the context of fire tests: for example 'stability' and 'integrity' were specific performance criteria under BS 476: Part 8[4].

A unified glossary of terminology relating to fire — principally fire tests — has been produced by ISO and the International Electrotechnical Commission[5].

Because there is concern that trade and advertising literature may unwittingly imply that products are safe in fire, eg *offering absolute safety,* BSI published a leaflet[6] directed at manufacturers (reissued in 1982[7] and 1988[8]). Another document, issued in liaison with BSI, is concerned with the publication of claims for the fire characteristics of products or systems[9].

In addition to giving examples of deprecated terminology, eg *fireproof* and *self-extinguishing* (which are 'coloured' terms), *incombustible* and *inflammable,* the BSI leaflet includes the following recommendations.

- Do not use terminology which:
  makes a claim which cannot be substantiated,
  may give a misleading impression of performance, or
  implies a judgement of performance in the event of a real fire.
- Substantiate your claims by quoting British Standard fire tests and results. Be sure that the test and test results upon which claims are based relate specifically to the material, product etc in question, and not to a similar or modified product. Be sure that the test reflects use of the product in practice.
- Be familiar with the definitions of terms associated with fire in BS 4422 and BS 6373[10], and use the correct terms.
- Avoid coining brand or trade names which may misleadingly imply fire safety.
- Avoid vague implications that the material complies with British Standards, Building Regulations, etc.
- If there are recognised fire hazards during the installation or use of a material, product, etc, warn the user and give advice on how to minimize risks.

In 1979 BSI published DD 64[11] (now BS 6336[12]) in which recommendations were made for a logical system of terminology. These recommendations arose out of concern felt over the possibility of confusion or misconception, leading even to danger, which lay in the terminology used. The guide attempts to define the limits within which fire tests can predict the hazard of any given fire situation and draws attention to the factors which should be considered by control or specifying authorities when specifying an acceptable degree of fire safety in the performance of a product or system.

Finally, it is stressed that although the expressions **fireproof, incombustible, inflammable** *and* **fire-resisting materials** *are quoted in Chapter 2 and Appendix 1, these expressions are historical and no longer used.*

Table 1 Fires by location, 1981–1991

Location	\multicolumn{11}{c}{Number of fires (thousands)[1]}
	1981	1982	1983	1984	1985	1986	1987	1988	1989	1990	1991
Total	**329.6**	**357.9**	**372.4**	**446.6**	**388.0**	**387.3**	**354.0**	**356.0**	**456.2**	**467.0**	**436.3**
Occupied buildings	**94.8**	**96.2**	**97.5**	**101.5**	**104.7**	**105.6**	**104.1**	**106.4**	**110.2**	**108.1**	**107.4**
Dwellings	56.0	56.4	57.4	59.0	62.6	63.5	63.2	64.2	64.5	63.2	64.1
Private garages, sheds, etc	6.2	6.9	6.7	7.6	6.9	7.0	6.8	6.9	8.0	8.3	7.6
Agricultural premises	2.5	2.5	2.3	2.4	2.4	2.5	2.2	2.0	2.2	2.1	1.9
Construction industry premises	1.0	1.0	0.9	1.0	0.9	0.9	0.8	0.7	0.2	0.6	0.6
Other industrial premises	7.5	7.8	7.6	8.3	8.3	7.9	7.8	6.9	7.0	6.5	5.8
Retail distribution	4.3	4.3	4.4	4.6	4.6	4.6	4.3	3.9	4.1	4.2	4.3
Hotels, boarding houses, hostels, etc	1.7	1.6	1.7	1.7	1.8	1.8	1.8	1.9	2.2	1.9	1.8
Restaurants, cafes, public houses, etc	2.1	2.0	2.1	2.2	2.3	2.3	2.4	3.1	3.2	3.3	3.4
Education	1.8	1.8	1.8	2.0	1.9	1.8	1.7	2.0	2.3	2.3	2.3
Hospitals	2.1	2.2	2.2	2.1	2.3	2.2	2.3	2.4	2.4	2.3	2.3
Recreational and other cultural services	1.3	1.3	1.5	1.6	1.4	1.5	1.4	1.4	1.6	1.5	1.6
Other and unspecified	8.4	8.3	8.8	9.0	9.4	9.6	9.4	10.8	12.4	11.9	11.7
Chimney fires	**44.1**	**43.9**	**46.9**	**50.5**	**63.8**	**54.0**	**48.6**	**40.6**	**37.2**	**33.1**	**39.0**
Outdoor fires[2]	**44.5**	**50.5**	**52.9**	**59.5**	**57.3**	**60.2**	**61.5**	**60.0**	**66.7**	**70.8**	**83.9**
Secondary fires[3]	**146.2**	**167.3**	**175.1**	**235.0**	**162.2**	**167.4**	**139.7**	**149.0**	**242.1**	**255.0**	**205.9**

1 Figures are rounded and the components do not necessarily sum to the independently rounded totals.
2 Includes road vehicles and derelict buildings.
3 Those involving only single derelict buildings, single buildings under demolition or such outdoor locations as grassland, railway embankments or refuse (except such fires which involve casualties, rescues or escapes, spread beyond the location of origin or are attended by five or more appliances, all of which arrived at the fire ground and were used in fighting the fire).

Table 2 Fires in dwellings and other occupied buildings by cause, 1981–1991

Number of fires (thousands)[1]

Cause	1981	1982	1983	1984	1985	1986	1987	1988	1989	1990	1991
Total — in all causes	56.0 (38.8)	56.4 (39.8)	57.4 (40.1)	59.0 (42.6)	62.6 (42.0)	63.5 (42.2)	63.2 (40.9)	64.2 (42.2)	64.5 (45.6)	63.2 (44.9)	64.1 (43.3)
Malicious[2]	4.5[3] (8.3)[3]	5.0 (8.9)	5.7 (9.8)	6.5 (11.3)	7.2 (11.2)	7.6 (11.7)	7.7 (11.1)	8.1 (11.8)	9.5 (13.3)	9.7 (13.9)	9.9 (15.0)
Accidental or unspecified causes	51.5 (30.5)	51.4 (30.9)	51.8 (30.2)	52.5 (31.3)	55.4 (30.8)	55.9 (30.4)	55.5 (29.8)	56.1 (30.4)	55.0 (32.4)	53.4 (31.1)	54.2 (28.4)
Faulty fuel supplies	2.2 (1.9)	2.1 (1.9)	2.5 (2.1)	2.1 (1.9)	2.2 (2.0)	2.3 (2.0)	2.4 (2.0)	2.1 (1.9)	2.0 (2.0)	2.1 (2.0)	2.1 (1.9)
Faulty leads to appliances	1.3 (1.0)	1.2 (1.0)	1.2 (0.9)	1.3 (1.0)	1.6 (1.1)	1.5 (1.1)	1.4 (1.1)	1.3 (1.1)	1.2 (1.0)	1.1 (1.1)	1.0 (1.0)
Faults in appliances	6.1 (4.6)	6.4 (4.9)	6.1 (4.7)	6.0 (4.8)	6.7 (5.4)	6.8 (5.0)	6.9 (5.3)	6.9 (5.8)	6.4 (5.8)	6.4 (5.4)	6.9 (5.2)
Misuse of equipment or appliances	23.7 (5.0)	23.0 (5.3)	23.5 (5.8)	24.3 (5.9)	25.5 (6.0)	25.9 (6.1)	26.7 (6.1)	27.8 (6.1)	27.3 (6.4)	26.3 (5.9)	27.0 (5.6)
Playing with fire	3.8 (3.8)	4.0 (3.8)	4.1 (3.5)	4.1 (3.6)	4.3 (3.4)	3.8 (2.9)	3.6 (2.7)	3.9 (3.0)	3.7 (2.8)	3.2 (2.0)	2.9 (1.7)
Careless handling of fire or hot substances	5.3 (4.0)	5.9 (4.2)	6.2 (4.2)	6.9 (4.5)	7.2 (4.3)	7.3 (4.3)	6.8 (4.2)	7.0 (4.3)	7.4 (4.5)	7.0 (4.3)	6.6 (3.8)
Placing articles too close to heat	3.7 (2.4)	3.7 (2.7)	3.5 (2.2)	3.4 (2.1)	3.7 (2.0)	3.9 (2.1)	3.8 (2.1)	3.6 (2.2)	3.2 (2.2)	3.2 (2.1)	3.4 (1.7)
Other	3.2 (4.3)	3.0 (4.1)	2.9 (4.2)	2.9 (4.9)	3.0 (4.6)	2.9 (4.6)	3.0 (4.6)	2.7 (4.3)	3.0 (6.0)	3.2 (6.8)	3.5 (6.1)
Unspecified	2.2 (3.6)	1.9 (3.1)	1.7 (2.8)	1.5 (2.6)	1.3 (2.1)	1.6 (2.3)	1.0 (1.8)	0.9 (1.7)	0.9 (1.6)	1.0 (1.4)	0.9 (1.3)

1 Figures are rounded and the components do not necessarily sum to the independently rounded totals.
2 Includes fires where malicious or deliberate ignition is merely suspected, and recorded by the brigade as 'doubtful'.
3 Excluding fires caused by terrorist activity.

NB Figures in brackets are those for 'Other occupied buildings'

Table 3 Fires in occupied buildings by spread of fire, 1981–1991

Year	Number of fires (thousands)[1] Total	Spread of fire Confined to room of origin	Beyond room of origin
1981	94.8	86.2	8.6
1982	96.2	87.7	8.5
1983	97.5	89.5	8.0
1984	101.5	92.1	9.4
1985	104.7	95.2	9.5
1986	105.6	95.8	9.9
1987	104.1	94.7	9.3
1988	106.4	96.7	9.7
1989	110.2	98.6	11.6
1990	108.1	95.6	12.5
1991	107.4	95.6	11.8

[1] Figures are rounded and the components do not necessarily sum to the independently rounded totals.

Table 4 Casualties from fires by location group, 1981–1991

Year	Fatal casualties Total	Dwellings	Other occupied buildings	Outdoors[1]	Non-fatal casualties Total	Dwellings	Other occupied buildings	Outdoors[1]
1981	975	780	80	115	8992	6343	1794	855
1982	919	728	80	111	9418	6659	1800	959
1983	903	710	93	100	10043	7137	1821	1085
1984	887	692	75	120	11116	7757	2155	1204
1985	978	700	105	173	11926	8501	2257	1168
1986	957	753	75	129	12768	9403	2200	1165
1987	929	710	92	127	12567	9476	1925	1166
1988	915	732	73	110	13376	10178	2188	1010
1989	901	642	77	182	14159	10388	2453	1318
1990	898	627	59	212	14041	10457	2301	1283
1991	816	608	59	149	14714	11215	2239	1260

[1] Includes road vehicles and derelict buildings.

Chapter 2: Historical background

Current statutory provisions within the United Kingdom have evolved from measures introduced slowly over many centuries. Early attempts were concerned mainly with minimising the possibility of outbreak of fire and its subsequent spread from building to building; measures largely brought about in reaction to major disasters. It was not until the nineteenth century that structural provisions were made for the safety of people within premises on fire — initially in connection with places of assembly and with factories.

Although large-scale tests were carried out in the mid-eighteenth century following the recognition that fire should be confined to the room of origin rather than the building, it was not until the end of the nineteenth century that we find the origins of standard fire test procedures.

Legislation

Middle Ages

Houses in mediaeval Britain were, except those of the wealthy, usually constructed with timber frames filled in with wattle and daub; the roofs were thatched and chimneys, as such, did not exist. Within the congested walled towns the houses were built in narrow streets with overhanging upper storeys, the top portions of which almost met those of the houses on the other side of the street. Any outbreak of fire could therefore spread very easily. With houses having domestic fires on a central hearth and the use of straw as a floor covering, together with the existence of thatched roofs, William the Conqueror ordained that all fires should be covered at night upon the ringing of a bell. The metal cover used for this purpose was called a Couvre Feu, which in course of time became Curfew. Nevertheless, over the next hundred years fires are recorded as having destroyed (or nearly so) London, Winchester, Worcester, Bath, Lincoln, Chichester, Rochester, Gloucester, Peterborough, York, Nottingham, Glastonbury and Carlisle (some more than once).

The first recorded attempt of any significance to legislate for the control of fire spread appears in the London Assize of 1189 by Henry Fitz-Ailwyn, Mayor of London. This laid down that houses in the City were to be built of stone, thatched roofs were not to be permitted, and party walls were to be of a minimum height and thickness. However, there was no means of enforcing control of these requirements.

A disastrous fire in London in 1212, where it is said that three thousand people died, led King John to issue an Ordinance which governed the construction of alehouses in the City. It also required that all wooden houses in Cheapside be pulled down or altered, no new roofs were permitted in thatch, and existing thatched roofs were likely to be demolished if not plastered over. Other requirements were made in connection with bakehouses and brewhouses. During the summer months houses were required to have a tub or tank of water standing ready in case of fire. In 1246 the Assize of 1189 was reissued and reaffirmed.

During the fourteenth century, a move away from centrally located hearths to a position against an outside wall began. However, it was not until the end of the century that chimneys came into use, but as they were generally constructed from hollowed-out logs they presented a greater danger than the methods of removing smoke previously adopted. Ordinances against the use of timber chimneys were passed the following century in London and Worcester. The fifteenth century also saw the first Act of Parliament relating to fire (James 1.1425. Chapter 23) which made provisions for fire prevention, fire fighting and penalties against persons causing fire in Scotland.

Seventeenth Century
Although building regulations of various kinds had existed in London for centuries (intensified by the Stuarts but without much success) London was still full of closely spaced wooden houses in 1666 — the poorer ones being back-to-back with pitch smeared weather boarding. The Great Fire of London started in the early hours of 2nd September at the end of a long dry summer, and after burning for four days had destroyed about 80% of the City.

London, which was used to dozens of minor fires yearly, at that time housed more than one-tenth of the population of the country and more than half of its wealth. With Britain at war with France, this fire was therefore a national as well as a local disaster — even though only six people died. The result was that London acquired its first complete code of building regulations and means for its implementation.

On September 13th Charles II issued a proclamation of which the basic principles were — the walls of all new buildings were to be constructed of brick or stone; the main streets were to be wide enough to prevent spread of flame; existing narrow alleyways were to be considerably reduced in number; and a survey was to be made of all ruins and ownership be shown of every plot. The King also appointed three Commissioners 'to join with such surveyors and artificers as the City might appoint' to press forward with the work of the survey. Three surveyors were appointed, who, with the Commissioners, managed the survey as well as devising building regulations as to the width of several classes of streets and the type of house that should be allowed in each. These regulations were embodied in 'An Act for Rebuilding the City of London'* which received Royal Assent on 8th February. The thickness of walls was related to their height; timber was not permitted on the exterior of houses (except for oak door-cases, window frames and bressummers, ie lintels, and for modifications necessary for shops); the

---
* Including Westminster.

proximity of timber to flues and chimneys was controlled; and details of permitted building construction were set out with provision for penalties for those who did not conform to them. Dangerous and noisome trades were also prohibited in the 'high' streets. Figure 2 shows three of the four permitted house types with required thicknesses of brick walls given in terms of numbers of bricks, eg 1½ br. (no schedule having been prescribed for Mansion Houses not fronting a street or lane, which also could not exceed four storeys — excluding cellars and garrets).

In 1684 a mutual Friendly Society was formed for assisting members in the event of fire — calculations for charges being based upon the assumption that one house in two hundred was burned down every fifteen years. Earlier attempts at fire insurance were made in 1635 and 1638 when certain London citizens had joined together in a pool to protect householders from losses arising from fire (at a rate of two percent on the yearly rent)[13].

Following a history of serious fires in Edinburgh, where in Mary's reign (under the French influence) the small wooden cottages covered with straw had begun to be replaced by houses of ten or twelve storeys fronting narrow streets, an 'Act Regulating the Manner of Building within the Town of Edinburgh' was passed in 1698 which required that in future no building should exceed five storeys. Rules of

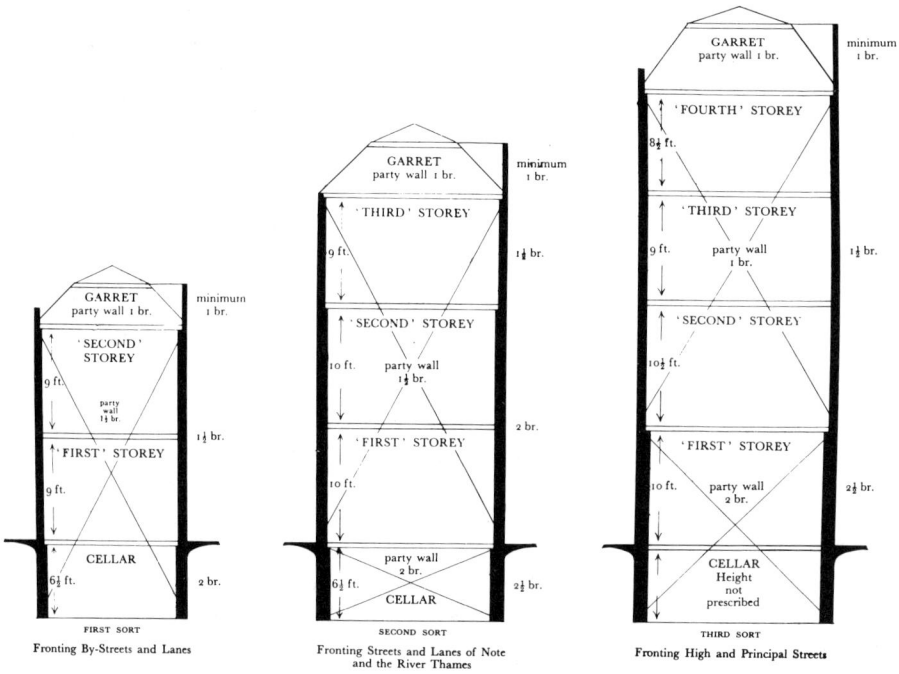

Figure 2 Sections of houses authorised by the Rebuilding Act of 1667

construction were given, including the thickness of party walls and the construction in and around chimneys. This Act reinforced an Edinburgh council statute of 1674 which in recognising the problems caused by timber frontages, narrow streets and high buildings had required that all new tenements be built in stone. As elsewhere, either the rules were not enforced or were ineffective.

Eighteenth Century
Throughout this century many fires occurred in London where over one hundred houses were destroyed on each occasion, and in the provinces many towns and villages were partially or completely destroyed by fire. In fact between 1600 and 1799 there were forty-one fires in which 100 houses or more were destroyed (eg 600 houses in Northampton in 1675, 337 houses in Blandford Forum in 1731 and 460 houses in Crediton in 1743) and thirty-seven fires where at least 50 dwellings were destroyed. Seven towns (Blandford Forum, Chudleigh, Northampton, Southwark, Tiverton, Wareham and Warwick) followed the example set by London in procuring an Act of Parliament for rebuilding[14].

Various minor Acts relating to London were passed between 1707 and 1772 under which the building of overhanging wooden eaves and cornices was prohibited; external and party walls were to be carried up as parapets; the requirements for the thicknesses of party walls and the building in of timbers were amended; and minimum dimensions were set down for chimneys, flues and hearths and for the relation of timber to them. Provisions were also made for parishes to keep fire engines and for the Fire Insurance Offices to employ watermen to fight fires. The Act of 1724 also extended the provisions of the Acts to St Pancras, St Marylebone, Paddington and Chelsea. These Acts were superseded by the Fires Prevention (Metropolis) Act of 1774 which lasted seventy years without amendment. This Act listed buildings into seven classes with thicknesses of external walls and party walls laid down for each class; required (for the first time) that proper party arches and floors be constructed in buildings of mixed user; and listed permissible materials for the construction of walls and roofs. The Act also included provisions with respect to the maximum area of warehouses, and for the appointment of Surveyors* for all the London boroughs and not just for the City as previously. No openings were allowed in party walls other than doorways between two warehouses, or between two stables, and these doorways were to be fitted with iron doors. The Act also decreed that every parish should provide 'three or more proper ladders, of one, two and three storey high, for assisting persons in houses on fire to escape therefrom'.

Theatres had been taking their toll of fire victims for many years when in 1794 the rebuilt Drury Lane Theatre incorporated an iron safety curtain in addition to a water tank situated on the roof which was intended for use in the event of fire as well as for aquatic scenery. Other eighteenth-century innovations were the treatment of theatrical scenery with alum, borax or ferrous sulphate, or its being covered with a mixture of plaster of paris and clay, and in 1786 Arfird suggested the use of ammonium phosphate for fireproofing theatre curtains. Meanwhile

---

* It was not until the 1844 Metropolitan Building Act that the term 'District Surveyor' was first used.

under the 1751 Disorderly Houses Act 'any house, room, garden or other place kept for publick dancing, musick or other publick entertainment of the like kind, in the cities of London and Westminster, or within twenty miles thereof, without a licence' was deemed a disorderly house or place.

The end of this century brought the introduction of hollow pot forms of construction and the first mill to use iron columns and beams to carry its 'fireproof' floors was erected in Shrewsbury.

During the eighteenth and nineteenth centuries the problem caused by the growth and expansion of towns led many municipal corporations to seek private Acts of Parliament to control sanitation, streets and buildings.

Nineteenth Century
Numerous fires throughout the eighteenth and early nineteenth century in Edinburgh, culminating with several fires in 1824, resulted in the formation of the Edinburgh Fire Engine Establishment in October 1824. In 1850 the Burgh Police Act required that party walls be carried through the roof and that all party walls, external walls and roofs be constructed in 'incombustible' materials. The Burgh Police Acts were generally adoptive and often followed the more progressive local authorities both in Scotland and England.

Liverpool obtained a Building Act in 1842 and Fire Prevention Acts in 1843 and 1844, but these apparently fell short of advanced opinion.

The 1844 Metropolitan Building Act, which was based on the Fires Prevention (Metropolis) Act of 1774, made few marked changes in constructional requirements for fire. However, the seven classes of buildings were reduced to three — namely dwelling houses, warehouses and public buildings. Warehouses were limited to 200,000 cubic feet without party or division walls, and the floors of all halls, corridors, stairs and landings of public buildings had to be 'fireproof'. Under the Metropolitan Building Act of 1855, the Metropolitan Board of Works (which was replaced in 1889 by the London County Council (LCC)) was empowered to alter the rules regarding the thicknesses of walls, etc. Building in iron or steel was permissible subject to the approval of the Board.

For the remainder of England and Wales, the 1847 Towns Improvement Clauses Act and the 1847 Town Police Clauses Act set out various standard clauses usually contained in local Improvement Acts, eg construction of roofs and walls, and the purchase of fire engines. In 1858 the Local Government Act gave urban authorities the power to make building byelaws subject to confirmation by the Home Office. However, in 1871 this responsibility was transferred to the Local Government Board who, with the Royal Institute of British Architects (RIBA), drew up Model Byelaws (first issued in 1877) owing to the wide divergencies between byelaws passed by different authorities. The original scope of the byelaws was extended under the Public Health Act of 1875 and the Public Health Acts Amendment Act of 1890. (Separate Model Byelaws for rural authorities were issued in 1901.)

In 1861 the Tooley Street warehouse fire (Figure 4), which cost the insurance companies over £2 million, led Captain Shaw of the London Fire Engine Establishment to state that a 216,000 cubic foot building was the largest volume that could be protected with reasonable hope of success. In February 1862 the insurance companies, who were responsible for the London Fire Engine Establishment, wrote to the Home Secretary saying that they could no longer be responsible for the safety of London from fire. The Select Committee set up 'to enquire into the existing state of legislation and of any existing arrangements for the protection of life and property against fires in the Metropolis' was told that the London Building Acts were faulty and inadequate, and were far inferior to the Bristol and Liverpool warehouse Acts. Apparently dock warehouses were exempt and one warehouse had been built with an undivided volume of 5,000,000 cubic feet. In several instances it had been successfully claimed that certain warehouses were not warehouses within the specific terms of the Acts.

A further Select Committee was set up in 1867 'to inquire into the existing legislative provisions for the protection of life and property against fires in the United Kingdom, and as to the best means to be adopted for ascertaining the causes, and preventing the frequency of fires'. They recommended that a general building Act for all towns in the United Kingdom should be placed on the Statute Book (with similar provisions and powers to the Metropolitan and Liverpool Building Acts). This should embrace thicknesses and heights of party walls; the placing of fireplaces, stoves and flues and the proximity of timber to them; the limiting of size and isolation of warehouses; and the use of proper materials in building. Other recommendations concerned living accommodation above shops; large lodging houses; water supplies; the classification and storage of goods; the sale and storage of flammable substances; and the setting up of procedures for investigating the cause of every fire[15]. Unfortunately no action was taken.

Legislation for theatres had existed outside London under the 1843 Theatres Act. However, the report which followed the inquest on the 1887 fire at the Theatre Royal Exeter, in which 186 people died, probably assisted in the passage of Section 36 of the Public Health Acts Amendment Act of 1890 which stated 'Every building which, after the adoption of this part of this Act in any urban district, is used as a place of public resort, shall, to the satisfaction of the urban authority, be substantially constructed and supplied with ample, safe, and convenient means of ingress and egress for the use of the public, regard being had to the purposes for which such building is intended to be used, and to the number of persons likely to be assembled at any one time therein'. But a Bill introduced in 1890 'for the further protection of life and property in the United Kingdom' to give County Councils power to appoint inspectors and to insist on proper fire escapes and alarm apparatus at all hotels, factories, theatres, schools, hospitals and workhouses, was not passed.

The nineteenth century brought considerable improvement in London's theatres. All theatres, other than 'patent'* theatres, had to be licenced under the 1843

---

* The Theatre Royal (Drury Lane) and the Royal Opera House (Covent Garden) which enjoyed letters patent direct from the Crown.

Figure 3 Fire escapes with fly ladders and chutes in action (circa 1850)

Figure 4 Tooley Street fire 1861

Theatres Act as amended by the Local Government Act (England and Wales) of 1888. Under the 1878 Metropolis Management and Building Acts (Amendment) Act, the Metropolitan Board of Works was authorised to cause alterations in existing theatres and music halls to be made in cases in which special danger from fire may result to the public, and authorised them to make regulations with respect to the position and structure of new theatres (and certain new music halls) for the protection of the public against danger from fires. These regulations[16] were first issued in 1879 and were revised from time to time. The 1882 Metropolitan Board of Works (Various Powers) Act empowered them to make requirements in respect of any place of public entertainment as regards means of exit, nature of fastenings and notices.

In 1897, 124 people died and over 200 were injured at a Paris Charity Bazaar where none of the exits was indicated and untreated velaria (canvas awnings) covered the whole of the underside of the roof. This led the LCC to instruct its theatres and music halls' committee to report what security was afforded by the existing law to the public against fire and panic in charity bazaars and similar gatherings. On hearing that its powers were very limited, the Council sought legislation to license any establishments to which the public were admitted.

The 1894 London Building Act, which consolidated and amended the earlier legislation, had laid down that only by special permission of the Council could a warehouse be built with a volume of over 250,000 cubic feet unless there was subdivision by brick walls so that no division thereof extended to more than 250,000 cubic feet, and that no building (other than a church or chapel) was to exceed 80 feet (exclusive of two storeys in the roof and of ornamental towers, turrets or other architectural features or decorations) without the consent of the Council. The Act also made rules concerning new public buildings which included regulations (based on the earlier 1879 theatre regulations) on the width of staircases and exits, and required that every new building exceeding sixty feet in height be provided on the storeys over 60 feet above the street level 'with such means of escape in the case of fire for the persons dwelling or employed therein as can be reasonably required under the circumstances of the case' and that no such storeys be occupied until the Council had issued a certificate that the provisions of this section had been complied with. It also provided a schedule of 'fire-resisting materials' (see Appendix 1).

In spite of building legislation, there were ten fires in London during the last decade of the century — each destroying many buildings. The worst, in the Cripplegate area on 19th November 1897, involved four acres of warehousing — two and a half acres being completely burnt out with total losses estimated at £1.25 million (an enormous sum at that time). The fire had spread from building to building across narrow alleys, streets and courts, and four thousand people (mostly women and girls) were put out of work as a result.

The end of the century saw the introduction of the first control of means of escape in case of fire from factories under the Factory and Workshop Act 1891 which concerned premises in which more than 40 persons were employed; workshops in which more than 40 persons were employed being added by the Factory and Workshop Act 1895.

Scotland had two basic Acts containing requirements for buildings, but the large burghs and cities retained their own requirements. The Burgh Police (Scotland) Act 1892 followed the general approach of the earlier Burgh Police Acts, but the Public Health (Scotland) Act 1897 adopted the principles of the Public Health Act 1875 and gave powers to authorities in landward (ie rural) areas to make byelaws to control certain aspects of building.

Twentieth Century
In 1900 Glasgow obtained a Building Regulations Act which required 'fire-resisting divisions' in buildings exceeding 200 square yards in area which were used in part for trade or manufacture and in part for domestic use. Detailed byelaws were made under this Act in 1909.

The press reports which followed a fire in Queen Victoria Street, London, on 9th June 1902 in which eight people were trapped on the top floor and many girls threw themselves to death on the pavement rather than being burnt to death, caused much checking to be done of local byelaws in the provinces, as the Queen Victoria Street premises was not a workshop as defined in the 1901 Factory and Workshop Act (which had consolidated, with amendments, the powers under the 1891 and 1895 Acts). In London, the LCC were given powers under the London Building Acts (Amendment) Act 1905 to require works to be done to certain existing buildings to facilitate escape. The Act also required plans to be deposited for new building work, and a new schedule of 'fire-resisting materials' was published (among changes made, were the addition for general purposes of 'any combination of concrete and steel or iron' and, for special purposes, of wired glass, and a reduction in the minimum thickness for doors from 2 in to 1 3/4 in).

Since 1858 many Acts had been passed which had amended or extended the powers of statutory bodies, and an attempt was made to consolidate and clarify the law by the 1907 Public Health Acts Amendment Act.

In 1921 a Royal Commission was appointed 'to enquire into the existing provisions for (1) the avoidance of loss from fire, including the regulations dealing with construction of buildings, dangerous processes and fire risks generally, the arrangements for enquiry and research and for furnishing information and advice to public authorities and others on matters relating to fire prevention, and (2) the extinction of outbreaks of fire, including the control, maintenance, organisation, equipment and training of fire brigades in Great Britain, and to report whether any, and if so what, changes are necessary, whether by statutory provision or otherwise, in order to secure the best possible protection of life and property against risks from fire, due regard being paid to considerations of economy as well as efficiency'. Their report[17] published in 1923 made a number of recommendations, including — the need to extend the scope of the 1901 Factory and Workshop Act; fire precautions for new and existing theatres and other places of entertainment; the provision of adequate means of escape from hotels, boarding houses, flats, shops and similar premises with sleeping accommodation on the upper floors; and the drawing up of model regulations.

Meanwhile, the London County Council (General Powers) Act 1908 had dealt with the cubical extent and uniting of buildings, and the London County Council (General Powers) Act 1909 covered requirements for steel framed buildings, as well as authorising the Council to make regulations (issued in 1916) to govern the use of reinforced concrete. Under the London Building Acts 1930–39 powers were granted to make by-laws and the first set issued in 1938 covered many of the constructional matters previously contained in the earlier Acts. The by-laws were revised in 1952, 1972 and 1979. Powers were also granted in respect of means of escape from certain new and existing buildings based on their height and use; such powers being virtually identical to those covered by Sections 59 and 60* of the 1936 Public Health Act under which the 1939 and 1952 Model Byelaws for England and Wales were made. However, these byelaws were not entirely satisfactory in that local authorities were not obliged to adopt them, and many did not. The 1936 Act was therefore amended by the 1961 Public Health Act to permit the making of one set of building regulations to replace the 1400 sets of local byelaws. The first building regulations for England and Wales were made in 1965, and their scope (originally based on the 1936 Public Health Act) was extended through the Fire Precautions Act 1971 to include means of escape in case of fire.

Model byelaws were first issued in Scotland in 1932, one set for burghs and one set for landward areas. The set for burghs contained requirements for means of escape from public buildings as well as requirements for structural fire protection. The Model byelaws series continued to form the basis of control in all except the large burghs and cities until the introduction of the Building Standards (Scotland) Regulations 1963 under the Building (Scotland) Act 1959.

The twentieth century has also seen the introduction of a range of fire legislation in connection with occupied buildings — in most cases forming part of an Act dealing with much broader issues.

The Home Office, in the introduction to its 'Manual of safety requirements in theatres and other places of public entertainment' (produced in 1934)[18] explained that its recommendations were based on experience of disasters at home and abroad. Nine fires were highlighted, two of which (in Exeter and Paris) have been referred to earlier. The other fires mentioned, in chronological order, were at the Ring Theatre Vienna where 450 people died (1881); the Iroquois Theatre Chicago where 566 people died (1903); the Empire Palace Theatre Edinburgh (1911); the Grand Assembly Rooms Leeds (1923); Drumcollagher Co Limerick where 50 people died out of a total of 150 (1926); the Glen Cinema Paisley where 70 children were suffocated and crushed (1929); and Coventry (1931). Those at

---

*Now Sections 71 and 72 of the Building Act 1984.

Edinburgh, Leeds and Coventry were included as examples of safe escape being made. At Edinburgh, where although ten of the performers and stage staff died as a result of the fire, the whole of the audience of about 3000 apparently cleared the building in just under 2½ minutes while the band played the National Anthem. As a direct result of the Paisley fire, an amending regulation was issued under the Cinematograph Act 1909.

As regards factories, the Factories Act 1937 considerably extended the requirements as to means of escape originally contained in the Factory and Workshop Act 1901 under which District Councils had been given powers in respect of premises where more than 40 persons were employed. Following a fire* in February 1956 at the Eastwood Mills, Keighley, in which eight people died, the whole question of providing fire alarms and adequate means of escape was reviewed by the Factory Inspectorate and a survey of 40,000 to 50,000 premises was carried out. The 1961 Factories Acts consolidated the fire provisions contained in the Factories Acts of 1937, 1948 and 1959 — the last of which had placed the responsibility for certifying means of escape on the fire authority and not the district council as provided for earlier.

Another notable fire* occurred in June 1960 at William Henderson and Sons' department store in Liverpool. Ten people were trapped in the fourth storey even though the brigade arrived within two minutes of being called. One man fell to his death from a ledge while assisting others to safety. This fire (the main cause of its rapid spread being the presence of suspended ceilings and unenclosed escalators) prompted the fire sections contained in the 1963 Offices, Shops and Railway Premises Act — the clauses being modelled closely on the 1961 Factories Act.

Also in 1961, during the proceedings on the 1961 Licensing Act, opportunity was taken to table an amendment with the intention of giving fire authorities greater powers over club premises following the Bolton Top Storey Club fire* in May of that year. In this fire 19 people died — fourteen bodies were found in the club room and five of the people who jumped, or fell, into the adjacent river below also died.

A fire* at the Rose and Crown Hotel, Saffron Walden on Boxing Day 1969, where 11 people died, a number of others were injured whilst escaping and 17 people were rescued by ladders, was one of a number of hotel fires which gave added impetus to the passing of the Fire Precautions Act in 1971. In 1972, hotels and boarding houses were the first premises to be designated as requiring a fire certificate under the Act.

---

* Reports of these fires have been published by the Fire Protection Association (FPA), as have comments on the introduction of the various statutes[19].

Provision of means of escape from fire from houses in multiple occupation had been dealt with under Section 16 of the 1961 Housing Act (now controlled under Part XI of the Housing Act 1985). However, following a number of serious fires, concern was expressed during the proceedings on the 1980 Housing Act about fire safety in hostels. As a result, an Order[20] was made in 1981 which provided that exercise by local authorities of their powers under the Housing Act should be mandatory for certain premises.

Following the fire* on 11th May 1985 at the Bradford City football ground where 56 people died and many were seriously injured, a committee was set up to inquire into the operation of the Safety of Sports Grounds Act 1975. The recommendations made in their final report[21] resulted in changes introduced in the Fire Safety and Safety of Places of Sport Act 1987 and the revision of the Home Office guide to Safety at Sports Grounds.

Perhaps one of the most important pieces of work carried out this century in connection with fire, was that undertaken by a Joint Committee† of the Building Research Board and the Fire Offices' Committee‡ (FOC). The Committee, after looking at evidence available from home and abroad (including experience gained during the Second World War) published their report in separate parts in 1946[2] and 1952[22]. The purpose of the report was to review the underlying principles of fire protection in buildings in the light of the knowledge at that time and to present the results in the form of recommendations. Such recommendations were 'not intended as a code for immediate application in practice, but as an exposition of the subject in the light of present-day knowledge, for the information of persons or bodies concerned in formulating rules for legal, insurance and other purposes'. Many of the Committee's recommendations are embodied in current thinking.

It is perhaps worth noting that apart from guidance prepared by the LCC, specific guidance in respect of means of escape and fire precautions had generally not been available before the last war; a fact which prompted the Building Industries National Council (BINC) to publish in 1945 a model code of requirements for application throughout the country and applicable to all types of buildings[23]. (A report 10 years earlier was concerned only with London[24]). The first British Standard Code of practice on 'Precautions against fire' was published in 1948 (Houses and flats of not more than two storeys[25]), with other parts following in 1962 (Flats and maisonettes over 80 feet in height[26]), 1968 (Shops and departmental stores[27] and Office buildings[28]), and 1971 (Flats and maisonettes in blocks over two storeys[29]). These codes have been superseded by BS 5588: Parts 1[30], 2[31] and 3[32].

---

* Reports of these fires have been published by the Fire Protection Association (FPA), as have comments on the introduction of the various statutes[19].

† It was originally set up in 1938 and was commonly referred to as the Fire Grading Committee.

‡ An association, formally constituted in 1868, of companies that transacted fire insurance business and which, as part of its activities, has taken a prominent part in the encouragement of fire protection. In 1986 it became part of the Loss Prevention Council (LPC).

Fire testing

The appalling losses in the big fires of the first half of the eighteenth century, an age of growing scientific advancement, turned the minds of inventors to 'fireproof' buildings.

In 1776 David Hartley, with others, remained unharmed in an upper room in a house in Putney Heath (London) while a fire burned below — the floor having been protected with his system of patented metal plates designed to provide a complete barrier to the passage of flame or draughts of air (based on an idea first patented by his father of the same name in 1722). This experiment, which was a gala event held on the 110th anniversary of the Great Fire of London, followed three earlier successful tests.

In 1777 Viscount Mahon (who later became Earl Stanhope) and others successfully occupied an upper room in a house in Chevening (Kent) as part of a series of tests to prove the effectiveness of his 'stucco' treatment between joists.

Figure 5 Hartley's house at Putney Heath

In 1791 a number of leading architects in London formed the Association of Architects (later to become the RIBA) from which a committee was formed in 1792 to 'consider the causes of the frequent fires within the limits of the Act of the 14th of George III entitled "An Act for the further and better regulation of

Buildings, and Party Walls &c &c "* and the best means of preventing the like in future'. The committee resolving 'that as good party walls are the means of confining a fire to one house, so other practicable and not expensive means might be devised, which confine a fire to one room in a house', they procured two new houses in Hans Place, Chelsea, in which they carried out tests to evaluate the methods proposed by Hartley and by Stanhope. Tests were also carried out on floors treated with Henry Wood's liquid designed to render wood 'incombustible'[†]. The committee's report[33], together with appendices describing their tests and a copy of a paper given by Stanhope in 1778 in which he described his method of securing buildings against fire, was produced in 1793 by Henry Holland who considered the Act of 1774 to be insufficient, unintelligible and the source of perpetual contention. The tests carried out by the committee apparently also led to the discovery that a fire in an enclosed room often burns quietly for a considerable time, during which it gradually heats the contents of the room to their ignition temperatures and so results in the phenomenon which is now termed 'flashover'.

As a result of the fire at the Ring Theatre, Vienna, exhaustive experiments were carried out in Austria in 1885 and 1895 (and later in 1905 following the Iroquois Theatre fire) which added much to the knowledge of the value of roof vents and different types of safety curtains.

The disastrous fires in London during the last decade of the nineteenth century culminating in the Cripplegate fire, again focused on the importance of providing buildings with adequate fire protection. In 1897 Edwin O Sachs, with the aid of a group of influential men, formed the British Fire Prevention Committee (BFPC) to advise architects and engineers in their choice of 'fire-resisting materials' and to assist public authorities in framing by-laws and regulations for buildings generally. (A similar body had been formed in the USA in 1895.) In its first year no less than 425 members joined the BFPC, this high number being attributed in part to the Cripplegate fire. Sachs, although only in his twenties, was recognised as a leading authority on the design of theatres who, in the course of his work, was appalled by the conflagrations in the United States theatres and by the fire risks to which they were exposed. In 1899 the Committee opened Europe's first fire testing station near St John's Wood Road station in London (but later moved their premises to Westbourne Lodge, Bayswater and then, in 1905, to North Bank Lodge, Regents Park). Earlier experimental tests had been carried out in the USA in Denver (1890), New York (1895) and Brooklyn (1896); and in Germany in Munich (1885), Berlin (1893) and Hamburg (1895).

An impetus was given to the study of the fire resistance of building materials and structural elements by a resolution passed at the 1901 International Fire Congress to the effect that 'serious investigation into the fire resistance of materials and systems of construction should be supported both by the Government and local

---

*Fires Prevention (Metropolis) Act, 1774.
† It should be noted that timber or any other combustible material cannot be made non-combustible by the application of treatments.

Figure 6 General view of the BFPC's testing station near Westbourne Park

authorities, as well as by such of the Technical Societies to whose members the result of investigations of this description is of importance in the practice of their respective professions'[34]. In 1903, the International Fire Prevention Congress, held in London, condemned the use of the term 'fire proof' as applied to buildings or their components, and recommended the use instead of 'fire-resisting'. It also confirmed the BFPC proposed universal standards of fire resistance (under which tables set down minimum test performances for fire-resisting floors and ceilings, partitions and single doors — see Appendix 1), and advocated the establishment of testing stations in each country to conduct tests by the same methods[35].

The BFPC continued their work until 1920, the year after Sachs died, carrying out tests for fire resistance as well as on fire alarms and firefighting equipment. The results of their tests, as well as papers on fire related matters, were published in 263 red books. However, although the Department of Scientific and Industrial Research (DSIR) took over the material for 31 red books which they published in 1920 and 1921, no steps were taken to continue testing, pending the outcome of the 1921 Royal Commission.

In 1925 the National Fire Brigades' Association (NFBA) took over the work of the BFPC but until the opening of their own testing station at Bromley-by-Bow (London) in 1929, tests were either carried out at the sponsor's premises or at fire stations[36]. The NFBA continued testing until the Second World War, publishing their results as red books which adopted the BFPC numbering.

As the BFPC investigations suffered from not having generally accepted test procedures for obtaining comparable results, the RIBA in 1929 requested the British Engineering Standards Association (which later became BSI) to set up a committee to define such terms as fire resistance, incombustibility and non-inflammability and to specify procedures for measuring them by test. In connection with this work, the newly formed Building Research Station (BRS) was asked to review the published accounts of the fire testing methods adopted in different countries, preparatory to the design of suitable test furnaces.

In 1932 BSI published BS 476[37] which also covered materials for flues, furnace-casings, hearths and similar purposes. The object of the standard being 'not only to enable architects, engineers, builders and others using materials and/or methods of construction to know where they stand, but also to provide standards whereby old and new materials may be accepted or rejected on definite grounds by the authorities concerned'.

In 1935 the FOC (who had had test facilities in the Manchester area since 1889) opened their new site at Boreham Wood following collaboration with BRS. Arrangements were made with the Government whereby the new facilities (one wall furnace, one floor furnace and one column furnace) were made available to BRS for a programme of tests to measure the fire resistance of traditional and non-proprietary forms of construction. This research was undertaken so as to enable a change to be made (in the form of building regulations) from the

Figure 7 A group of the members of the 1903 International Fire Prevention Congress (Sachs seated fourth from right)

approach of specifying materials and thickness to a more functional type which could be framed in terms of a performance standard. The results of these tests carried out between 1935 and 1946 were published in 1953[34]. BRS was also responsible for carrying out sponsored tests to BS 476. In 1946 the Joint Fire Research Organization (JFRO) was formed as a partnership between the FOC and the Government (DSIR) and continued until 1976 when it was dissolved following a decision that the Government should no longer be involved in sponsored testing.

A test for surface spread of flame was added to BS 476 in 1945 following a programme of research which had been set up when it was realised that the standard test for combustibility/inflammability had failed to provide the necessary discrimination between internal lining materials. In 1953 the standard was revised[38] (with the tests for non-inflammability and for materials for flues, etc deleted) and later reissued as BS 476: Part 1, as a flammability test for thin flexible materials had been published separately as Part 2[39]. In the tests for fire resistance the method of expressing results by letter, ie Grade A (6h), B (4h), C (2h), D (1h) and E (1/2h) was discontinued and a three-hour classification was added. Other changes in connection with the testing of fire-resisting elements of structure were the deletion of the requirements for (i) the water jet test applicable to specimens tested for Grades A, B or C, and (ii) the impact test (a swinging 10lb solid cast-iron ball) on non-loadbearing specimens, excluding glazing, applied at 30 minutes and after the water jet test.

With the publication of separate test standards for non-combustibility (Part 4[40]), surface spread of flame (Part 7[41]) and fire resistance (Part 8[4]), BS 476: Part 1 was withdrawn (but remained available for sale as it was referred to in regulations).

In 1955 the first experiments were conducted with the fire propagation test apparatus (Part 6[42]) which had been developed following tests in 1949 in full-scale houses (and later at 1/2, 1/5 and 1/10 scale) which had shown that Class 1 in the surface spread of flame test included materials with a wide range of times to flashover[43]. Although BRS started work on the development of an external fire exposure roof test in 1936, work was suspended on the outbreak of war[44]. Development started again after the war and the test method (Part 3[45]) was published by BSI in 1958.

Bibliography

- **Anon.** An account of Great Fires which have happened in Edinburgh from 1700 to the late awful burnings. Edinburgh, T & W Nelson, 1824.
- **Chambers, Robert.** Notices of the most remarkable fires in Edinburgh from 1385 to 1824. Edinburgh, Chas Smith & Co, 1824.
- **Dyson T G.** A short history of the British Fire Service. International Fire Congress and Exhibition. London, June 1893.
- **Sachs, Edwin O.** Facts on Fires Prevention. Volume 1. London, B T Batsford, 1902.
- **The British Fire Prevention Committee.** First International Fire Prevention Congress. The Official Congress Report. London, BFPC, 1903.
- **Fletcher B.** The London Building Acts (Fifth Edition). London, B T Batsford, 1914.
- **Reddaway T F.** The rebuilding of London after the Great Fire. London, Jonathan Cape, 1940.
- **Gamble S G.** A Practical Treatise on Outbreaks of Fire (Third Edition). London, Charles Griffin and Company Ltd, 1941.
- **Wood, Marguerite.** All the statelie buildings of Thomas Robertson — a building speculator of the seventeenth century. The Book of the Old Edinburgh Club. Vol 24. Edinburgh, 1942.
- **Chanter H R.** London Building Law. London, B T Batsford, 1946.
- **Wood, Marguerite** (Editor). Extracts from the Records of the Burgh of Edinburgh 1665–1680. Edinburgh, Oliver and Boyd, 1950.
- **Brend H J.** Means of escape in case of fire. London, Sir Isaac Pitman & Sons Ltd, 1952.
- **Blackstone G V.** A history of the British Fire Service. London, Routledge and Kegan Paul, 1957.
- **Hamilton S B.** A short history of the structural fire protection of buildings, particularly in England. National Building Studies. Special Report No 27. London, HMSO, 1958.
- **Munro J R.** A history of fire prevention in buildings. The Architect and Surveyor. May/June 1970 and July/August 1970.
- **Knowles C C and Pitt P H.** The history of building regulation in London 1189–1972. London, Architectural Press, 1972.

Chapter 3: Existing legislation and other requirements

It has been generally accepted that the primary purpose of statutory requirements for structural fire precautions is to safeguard life; whereas the protection of property is assumed to be covered by insurance interests.

Legislation

Within the United Kingdom separate legislation exists for the construction of new buildings and for the control of fire precautions in occupied premises. In fact, thirty-nine statutory provisions relating to fire were listed in the 'Review of fire policy' published by the Home Office in 1980[46] (excluding the London Building Acts, Acts of Parliament applicable to some local authorities*, and provisions in respect of Northern Ireland, the Isle of Man and the Channel Islands).

The method of specifying legislative requirements depends upon the method adopted by, or permissible under, the rules which specify control and the legal status of the documents produced. Three methods are available and often a mixed system is employed. In a —

> *functional* system, the aims or objectives are specified and the designer or architect can, by reference to explanatory notes and guides or codes, provide evidence that the objectives are being achieved.

> *performance* based system, the objectives are translated into specific performance levels to be achieved in appropriate tests or evaluation techniques.

> *prescriptive* system, precise details of construction or permissible systems are given, which may possibly be specified on a deemed-to-satisfy basis.

Building Regulations
Separate regulations apply in England and Wales, Scotland, and in Northern Ireland, ie:

> The Building Regulations
> The Building Standards (Scotland) Regulations
> The Building Regulations (Northern Ireland)

Other statutory provisions apply within the Isle of Man and the Channel Islands.

---

*Information concerning provisions in local Acts is given by Pitt[47].

The scope of the different sets of regulations varies in accordance with the powers granted under the respective enabling legislation, but in general their objectives are the same although the method of control and manner of specifying requirements may vary.

Provisions for the control of structural fire precautions in connection with the proposed erection or alteration of a building (or a change of use) are 'designed to set down basic requirements for a building structure that will not itself be unduly hazardous in the event of a fire and which will, when coupled with satisfactory means of escape, ensure within reasonable limits the safety of all persons who might become involved'[48]. Therefore if a fire breaks out, it does not spread in a manner that would jeopardise the people in the building in the early stages of the fire, or lead to the structural collapse and possible conflagration. Other provisions are concerned with the construction and siting of chimneys, flues, hearths and associated appliances.

Thus requirements made in connection with regulations generally provide for the fire resistance of elements of building construction; the construction of external walls and nature of cladding; the nature of roof coverings; the restriction of spread of flame over surfaces of walls and ceilings; the sub-division (where appropriate) of buildings into separate fire compartments; and the construction of separating (party) walls between buildings.

In 1985 the Building Regulations for England and Wales changed to a system of functional requirements expressed in terms of 'adequate' and 'reasonable', with practical guidance on ways of meeting them given in a set of documents approved by the Secretary of State (Approved Documents). Means of escape in case of fire was dealt with differently by setting out mandatory rules[49]. Since then the Department of the Environment has reviewed the regulatory requirements and the technical content of the supporting documents. In connection with the fire safety provisions of the Regulations (Part B (Fire)), FRS commissioned a study[50] to examine and consolidate the practical information available both in the UK and overseas since the Fire Grading Report[2,22], and to propose changes that could be made without reducing the level of life safety. The revised Building Regulations[51] and Approved Document B[52] were issued at the end of 1991.

The Scottish Office introduced a 'functional' system of building regulations in 1990[53] with mandatory Technical Standards supported by deemed-to-satisfy provisions[54]; Northern Ireland are in the process of changing their building regulations[55] to functional requirements supported by deemed-to-satisfy Technical Booklets.

Within Inner London (see Figure 8), the various requirements made under the London Building Acts and By-laws in connection with the control of structural fire precautions were generally repealed by the Building (Inner London) Regulations 1985[56] which came into force on 6 January 1986. However, those powers which are not at present covered by the Building Regulations remain. For example, under Section 20 of the London Building Acts (Amendment) Act 1939 (which concerns buildings of excess height and/or additional cubical extent), the local council (after consulting the fire authority) may also impose conditions in connection with 'special fire risk areas'; the provision and maintenance of fire alarm, detection and extinguishing appliances/equipment/systems; effective means of removing smoke in case of fire; and adequate means of access for fire brigade personnel and appliances.

Figure 8 Greater London

In connection with the adoption by the member states of the European Community of the Single European Act which places obligations to the free exchange of goods and services by the end of 1992, Council Directive 89/106/EEC[57] sets out essential safety requirements relating to construction products. This Directive has been implemented in the United Kingdom by The Construction Products Regulations 1991[58].

As part of the Government's initiatives to lighten the burden on industry and commerce, the Department of Trade and Industry's Enterprise and Deregulation Unit commissioned a study[59] to review the interrelation of the Building Regulations and fire certification requirements and the way they are put into effect. This report identified a need for two comprehensive and complementary documents — one a national guide to the legislation and procedures, the other a design guide to fire safety in building, capable of forming a basic educational resource for professionals in the field. The first of these[60] explains the steps involved in approving fire safety aspects of building work, and the interaction between building regulations and other fire safety requirements in England and Wales.

Requirements in connection with occupied premises
In addition to the building regulations, fire precautions in occupied premises are dealt with under several separate forms of legislation. Table 5 shows a number of statutory provisions relating to means of escape in case of fire. However, it should be noted that:

- the extent of control varies widely,

- more than one statute may apply to a particular use, eg a cinema or hotel serving alcoholic drinks will also require a liquor licence, and

- in only a few cases have specific regulations been issued or guidance published.

The table does not identify premises due to be controlled under the proposed Fire Precautions (Places of Work) Regulations.

Fire Precautions Act 1971
Of the statutory provisions listed in Table 5, the most notable is the Fire Precautions Act 1971 which provides powers in respect of virtually all types of premises* in England, Wales and Scotland. In this connection it should be noted that a number of important amendments and additions have been made to the Act. Those changes made under the Fire Safety and Safety of Places of Sport Act 1987 are being implemented by a programme of phased commencement orders. Two such orders have been made so far — the first[61], which came into effect on 1 January 1988, concerned those matters not requiring supporting material; and the second[62], which came into effect on 1 April 1989, included the bringing into force of Sections† 5A, 9A and 9B.

Premises which require a fire certificate
The first designation order was made in 1972 for hotels and boarding houses[63,64]. Factories‡, offices, shops and railway premises§ followed in 1976[65] ¶ after section 1 of the Act had been extended to include 'use as a place of work' by the Health and Safety at Work etc Act 1974.
A fire certificate is required for —

1. hotels and boarding houses having sleeping accommodation for more than six persons (whether guests or staff), or where there is sleeping accommodation above the first floor or below the ground floor;

2. offices, shops and railway premises in which more than twenty persons are employed to work at any one time, or more than ten at any one time elsewhere than on the ground floor. **NB** In a multi-occupied building comprising several small premises, a fire certificate is required if the total number of employees meets these criteria;

---

* 'premises' means building or part of a building.
† of the Fire Precautions Act.
‡ Formerly controlled under the Factories Act 1961.
§ Formerly controlled under the Offices, Shops and Railway Premises Act 1963.
¶ This was superseded in 1989[66].

Table 5 Statutory provisions relating to occupied premises

	Celluloid and Cinematograph Film Act 1922	Petroleum (Consolidation) Act 1928	Licensing Act 1964	Fire Precautions Act 1971 (as amended)
Area applicable to	EWS	EWS	EW	EWS
Certificate or licence (where required)		L	L	C
Enforcing authority	LA	FA/LA‡	LJ	FA
Churches/chapels				
Cinemas	●			
Clubs Licensed/Registered			●	
Common lodging houses				
Flats/tenements				
Garages and certain car parks		●		
Homes Children's and nursing				
Hotels and boarding houses				●
Hospitals				
Houses in multiple occupation				
Industrial General Large chemical works (eg major hazards)				●
Inns				
Licensed premises (for the sale and supply of intoxicating liquor)			●	

NOTE: This table is not intended to be exhaustive and is only provided as a guide. In all circumstances a check should be made with the local authority regarding the legislation applicable to any particular scheme.

Health and Safety at Work etc Act 1974 (Fire Certificates (Special Premises) Regulations 1976)	Licensing (Scotland) Act 1976	Building Act 1984 Sections 71 and 72†	Fire Services (Northern Ireland) Order 1984 (designated use)	Cinemas Act 1985	Housing Act 1985 (as amended)	Housing (Scotland) Act 1987
ESW	S	EW	NI	EWS	EW	S
C	L		C	L		
HSE	LJ	LA	FA	LA	LA	LA
		●				
				●		
	●					
					●	
		●				
		●				
		●	●			
		●				
		●			●	●
●			●			
		●				
	●					

continued overleaf

Table 5 (continued)

	Celluloid and Cinematograph Film Act 1922	Petroleum (Consolidation) Act 1928	Education Act 1944 (as amended)	London Government Act 1963 Schedule 12	Private Places of Entertainment (Licensing) Act 1967	Theatres Act 1968
Area applicable to	EWS	EWS	EW	GL	EW	EWS
Certificate or licence (where required)		L			L	L
Enforcing authority	LA	FA/LA‡	LA	LA	LA/LJ	LA
Music and dancing, etc				●	●	
Offices						
Places of public Amusement Entertainment Resort Worship				●		
Restaurants						
Railway premises						
Schools			●			
Shops						
Sports grounds Ground (accommodation >10 000) Covered stand (accommodation ≥ 500)						
Stores/warehouses						
Storage of Celluloid Petroleum spirit, etc	●	●				
Theatres						●

KEY:
EW — England and Wales
EWS — England, Wales and Scotland
S — Scotland
NI — Northern Ireland
GL — Greater London

FA — Fire Authority
LA — Local Authority (Building, Education or Housing)
LJ — Licensing Justices, Magistrates, or Sheriff
HSE — Health and Safety Executive
SED — Scottish Education Department

Fire Precautions Act 1971 (as amended)	Safety of Sports Grounds Act 1975 (as amended)	Education (Scotland) Acts 1980 and 1981	Local Government (Miscellaneous Provisions) Act 1982 Schedule 1	Civic Government (Scotland) Act 1982 Schedule 1	Building Act 1984 Sections 71 and 72[†]	Fire Services (Northern Ireland) Order 1984 (designated use)	Fire Safety and Safety of Places of Sport Act 1987 (Part III)
ESW	EWS	S	EW*	S	EW	NI	EWS
C	C		L	L	EW	C	C
FA	LA	SED	LA	LA	LA	FA	LA
			●				
●						●	
			●	● ●	● ●		
					●		
●							
		●			●		
●					●	●	
	●						
							●
					●		
					●		

KEY (continued)

* — Excluding Greater London   † — Excluding Inner London

‡ — HSE is the enforcing authority at sites subject to the Notification of Installations Handling Hazardous Substances Regulations 1982

NOTE: In connection with premises coming within the provisions of the Fire Certificates (Special Premises) Regulations 1976, and Sections 71 and 72 of the Building Act 1984, see Appendix 2.

3   factories in which the number of employees conforms to the criteria in item 2 above, or where explosive or highly flammable materials are stored or used in or under the premises (**other** than such quantities of materials which the fire authority does not consider as constituting a serious additional risk to persons in the premises in case of fire).

As each designation order has been made, the Home Office and Scottish Office have issued a guide under which Part I sets down the provisions of the Act as it applies to the premises in question, and Part II is a guide to the basic standards for means of escape in case of fire and other related fire precautions. Also, when any class of premises is designated from the list given in section 1 of the Act (see Appendix 2) any relevant existing legislation for that class and covering the same range of fire precautions is repealed.

Revised guides have been published for:

- Premises used as hotels and boarding houses which require a fire certificate[67]
- Existing places of work that require a fire certificate[68]

which supersede the Guides to the Fire Precautions Act 1971 Nos 1, 2 and 3[69,70,71]. Separate guidance has been issued for managers of hotels[72] and places of work[73].

For those premises requiring a fire certificate under the Act (Section 5), such a certificate is issued by the fire authority when they are satisfied that the following matters are such as may reasonably be required in the circumstances for that particular use of the premises:

(a)   the means of escape in case of fire,
(b)   the means (other than means for fighting fire) for securing that the means of escape can be safely and effectively used at all material times,
(c)   the means for fighting fire [for use by persons in the building]*, and
(d)   the means for giving warning in case of fire.

Section 6 of the Act specifies the contents of fire certificates (see Appendix 2).

The Act also requires (Section 8) that the fire authority be informed if it is proposed to make a material extension of, or material structural alteration to, a certificated premises; or to make a material alteration in the internal arrangement of the premises or in the furniture or equipment with which the premises are provided; or (in the case of factory premises) to begin to store or use explosive or highly flammable materials in the premises, or materially to increase the extent of such storage or use.

The expression *material* is not defined in the Act. However, the new guide for hotels and boarding houses explains that 'it is considered that an alteration is material if it would render the means of escape and related fire precautions

---

* The wording in square brackets will be deleted on 1 August 1993 (SI 1993 No 1411) so as to enable sprinklers or other automatic extinguishing systems to be required as a condition of issuing a certificate.

inadequate in relation to the normal conditions of the use of the premises at the time the fire certificate was issued. It is unlikely, therefore, that the fire authority will need to be informed each time it is proposed to redecorate (unless the redecoration affects the means of escape), but there is an obligation to do so if the proposals involve structural alterations to the means of escape and its associated matters. In case of doubt the fire authority should be consulted'.

The fire authority deals with general fire precautions within all premises covered by the Fire Precautions Act. In the case of Crown premises, whether owned or occupied by the Crown (or both), the enforcing authority is Her Majesty's Inspector of Fire Services. (In this connection, the local fire authority is now responsible for National Health Service premises following removal of Crown immunity on 1 April 1991[74].)

It should also be noted that under the Fire Precautions Act a duty is placed on the:

- Local Building Control Authority in England and Wales to consult the fire authority (Section 16) before passing plans in connection with the erection or structural alteration of a building, or change of use, which on occupation would be a designated use.

- Fire authority to consult with the local authority (Section 17) before requiring alterations to be carried out prior to the issue of a fire certificate. However, the fire authority cannot make such requirements (Sections 13 and 14) if the plans of the building comply with the building regulations, unless:

    (i) they consider that the means of escape are inadequate for any reason which was not required or disclosed under the building regulations; or

    (ii) it is necessary to make requirements in order to satisfy any regulations made under Section 12.

As regards Northern Ireland, the Fire Services Order of 1984[75] includes similar provisions to those contained in the Fire Precautions Act. So far, designating orders have been made in respect of hotels and boarding houses[76]; factory, office and shop premises[77]; leisure premises[78]; and betting, gaming and amusement premises[79].

<u>Premises which do not require a fire certificate</u>
When any class of premises is designated from the list given in section 1 of the Act, a fire certificate is not necessarily required for all premises falling within that particular class. For example, any hotels, boarding houses, offices, shops, railway premises and factories which do not meet the relevant criteria listed under items 1, 2 and 3 on pages 31 and 36, are not required to have a fire certificate.

Apart from the general exemptions from the need to have a fire certificate set out above, the fire authority has powers under Section 5A to grant exemption in particular cases. A description of the types of premises that might qualify for exemption is given in The Fire Precautions (Factories, Offices, Shops and Railway

Premises) Order 1989[66]. However, there is a general duty under Section 9A to provide reasonable means of escape and means for fighting fire in these premises, and a code of practice[80] has been issued which gives practical guidance on how to comply with this duty in places of work. This code replaced the Fire Precautions (Non-Certificated Factory, Office, Shop and Railway Premises) Regulations 1976[81].

Regulations[82] have also been issued in Northern Ireland which require certain fire precautions to be taken in factories, office and shop premises which do not require a fire certificate.

Under Section 12 of the Act the Secretary of State may make regulations dealing with fire precautions. Such regulations have so far been made in respect of non-certificated factory, office, shop and railway premises[81] and, following the fire at King's Cross in 1987, sub-surface railway stations[83]. (Section 12 is reproduced under item B of Appendix 2.)

It is also proposed to make regulations under Section 12 to implement the general fire safety provisions of two European Council Directives relating to health and safety at work[84,85], in respect of fire safety in premises. The proposed regulations (the Fire Precautions (Places of Work) Regulations [1993]) will apply to premises which are used as a place of work by one or more employees. They will therefore not cover premises in which only self-employed persons are present. Exemptions from the regulations are likely where adequate alternative provisions can be demonstrated.

Two guides are to be published. One will be a simplified guide which should be sufficient for most employers, and the other a technical guide both for architects, designers, fire authorities and professionals in the field of fire precautions, and for employers who require more technical detail.

It is also proposed to produce regulations in due course in respect of fire precautions for moored vessels, tents and other movable structures, and places of work in the open air.

Health and Safety at Work etc Act 1974
It is the Health and Safety Executive (HSE) which is responsible for 'process' fire precautions within factories (ie the precautions taken to prevent an outbreak of fire and to minimise the amount of available fuel if there should be), and for 'Special Premises' which are dealt with under separate powers, namely the Fire Certificates (Special Premises) Regulations 1976[86]. (Similar regulations have been made for Northern Ireland[87].) The premises listed under item C in Appendix 2 require a fire certificate under these regulations. Such certificates specify broadly similar provisions and are able to impose similar requirements to those mentioned in respect of certificates issued under the Fire Precautions Act.

The HSE explain in their leaflet '20 Questions'[88] that the reasons why the issue of a fire certificate for certain premises rests with them rather than the local fire authority are:

(a)   In some premises the process hazards may be on such a scale or of such a character or have such a direct bearing on general fire precautions that the two aspects cannot be considered apart.
Items 1 to 9 in Appendix 2, Section C, belong in this category.

(b)   In certain premises where highly specialised processes are carried on (such as nuclear installations, explosives factories and mines) it is desirable that the HSE, which deploys the specialist inspectorates responsible for safety at such premises, should retain the fire certification responsibility.
Items 10 to 14 in Appendix 2, Section C, belong in this category.

(c)   Fire certification at temporary site buildings used as offices or workshops at building operations and works of engineering construction presents certain problems arising mainly from rapidly-changing conditions. HSE inspectors already visit building sites for other purposes and by including these premises in the regulations the HSE becomes responsible for fire precautions at them.

The leaflet points out that although many of the classes of premises listed are defined by reference to the use or storage of particular materials in specified quantities, the substances and quantities are not chosen because they are dangerous in themselves but because they give an *indication* that processes are undertaken of such a type, or on such a scale, as to warrant special consideration of the general fire precautions. The HSE have issued guidance on general fire precautions in such premises[89].

Housing Act 1985
As regards houses in multiple occupation*, local housing authorities in England and Wales have power under section 352 of the Housing Act 1985 (as amended by the Local Government and Housing Act 1989) to require the provision of adequate means of escape from fire and adequate other fire precautions (such as firefighting equipment and fire detection and warning systems) for the number of

---

* Houses and flats occupied by persons who do not form a single household (including, for example, hostels and lodging houses).

occupants. In the case of houses in multiple occupation which comprise at least three storeys (excluding basements) with a combined floor area of all storeys in excess of 500 square metres, the exercise of this power is mandatory in respect of means of escape from fire. Joint circulars giving guidance to local housing and fire authorities on the scope and nature of their powers and responsibilities relating to houses in multiple occupation were issued in 1986[90] and 1990[91], and guidance on the application of section 352 was published in 1992[92]*. Separate guidance[94] has been issued by the Scottish Office in connection with the Housing (Scotland) Act 1987.

Fire insurance requirements

As well as satisfying the foregoing statutory requirements in connection with new buildings and occupied premises, consideration may also need to be given as to whether the proposals meet with insurance requirements. Under the code of practice[95] published by the Loss Prevention Council (LPC)[†] it may be necessary for the building to satisfy a higher, or different/additional standard of construction.

The fire insurers' requirements contained in the code of practice are generally applied to buildings of all occupancies, particularly industrial and commercial, but specifically exclude domestic dwellings. The 'Standards of Construction'[‡] (now withdrawn) were generally applied to buildings erected prior to 1 March 1978. The 'Rules for the Construction of Buildings, Grades 1 and 2' (now withdrawn) were generally applied to buildings erected after March 1978.

Incorporated in the code of practice are the 'Rules for the Construction and Installation of Firebreak Doors and Shutters'. A list of those doors and shutters accepted by the fire insurers is contained in the 'List of Approved Products and Services' published annually in March by the LPC.

Buildings conforming to the requirements of the code of practice will not incur any additional premium. However, reduced levels of protection may be acceptable in some circumstances, but it is important that approval and advice should be sought from the insurers at the planning stage.

It should also be noted that since buildings, ranges of buildings or parts of buildings are normally rated as one risk **unless** separated by *fire break walls* and *fire break floors,* this could have design implications.

---

* This guidance replaces that published in 1988[93].
† Incorporating the Fire Offices' Committee.
‡ Rules for the Construction of Buildings. Standards I – V.

Chapter 4: Understanding fire

A designer of a structure will need to know a great deal about the conditions that will be imposed on that structure in service. Factors such as loading, environment and durability all have to be understood and assimilated into the design process. In the same way, when considering the provision of fire protection in buildings as part of a design process, it is axiomatic that a knowledge and understanding of fire behaviour is acquired so that fire protection principles can be soundly applied.

Unfortunately many obstacles confront the would-be student of fire. Researchers working under carefully controlled conditions in laboratories often have the greatest difficulty in producing repeatable fires for experimental purposes. It is often said that no two fires are alike. This is a reflection of the complexity of fire phenomena and the fact that the interaction of a fire with its environment is very much a probabilistic process. Anyone familiar with the domestic chore of fire-lighting will be well acquainted with the vagaries of fire.

In spite of these difficulties we can, by the adoption of simple models, gain an insight into the process controlling the growth and development of fires. It is also possible to make estimates of fire severity in a manner related to the performance of the building structure.

Burning is often defined as a set of complex chemical reactions in which substances combine with oxygen, producing heat and light. Fortunately, for most materials, these reactions do not take place readily; they can however be initiated by external stimuli such as raising the temperature. When, after removal of the source, the heat generated by the reaction is sufficient to cause yet more of the material to react with the further evolution of heat, we refer to the process as sustained ignition. Under favourable conditions, once ignited a fire will continue to grow until its size is controlled by the availability of fuel or by the supply of oxygen needed for combustion. It is useful to consider a fire in a compartment (ie room or other enclosure) as having three distinct phases:

(a) *The growth period* — initially we will be concerned with the ignition of materials, usually from a small source. Local flaming can develop and the feedback of generated heat may lead to the propagation of flame.

(b) *Steady state combustion* — the fire cannot keep growing indefinitely and its development continues until severity is controlled by factors such as compartment configuration, fuel availability and the degree of ventilation.

(c) **Decay** — as the fuel supply is depleted the fire will reduce in severity and eventually die down. It is expected that extinguishment will have taken place before this stage is reached.

These ideas can be expressed in a simple diagram (Figure 9).

We have already drawn attention to the unpredictability of fire. At this stage, in our simple model, no reference has been made to specific time scales or to the maximum temperatures that might be achieved. The first of these is rather indeterminate and the latter has perhaps less significance than we might imagine. From the fire protection point of view it is however necessary to distinguish clearly between phases (a) and (b).

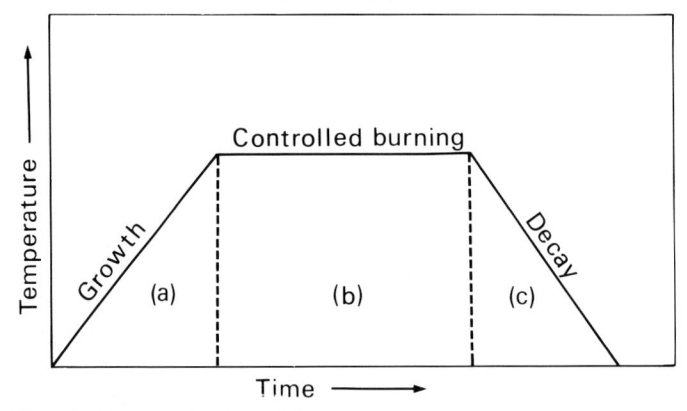

Figure 9 Fire — a simple model

The growth of a fire
Ignition
We have already referred to ignition as the process preceding burning. Ignition involves the bringing together of a source of heat and a suitable fuel in such a way that a continuing combustion reaction results. Materials vary in their susceptibility to ignition and even these initial stages of fire are very complex.

The first requirement is that the fuel itself needs to be raised to some temperature, T, so that given a supply of oxygen the combustion reactions take place readily and are self-sustaining. Again when attempting to simplify and break down the model it is found that each level has its own layers of complexity. If we attempt to measure ignition temperatures the values obtained are a function of the way in which the measurement is made and on how ignition has been defined. Because of this, there is a multitude of terms such as flash point, fire point, ignition temperature and auto ignition temperature, each of which in some way is meant to represent the ease with which materials can be set on fire. Usually confusion can be resolved by reference to specific Codes or to Standards such as BS 4422 'Glossary of terms associated with fire'.

Notwithstanding these difficulties we will assume that an ignition temperature for a material can be defined and exists. Table 6 gives some values for ignition temperatures published by Hilado[96].

Table 6 Ignition temperatures

Material	Flash ignition Temperature °C	Self ignition Temperature °C
Cotton	230 – 266	254
Paper	230	230
White Pine	228 – 264	260
Douglas Fir	260	
Wool	200	
Polyethylene	341	349
PVC	391	454
Polymethyl Methacrylate (Perspex)	280 – 300	450 – 462
Polystyrene foam	346	491
Polyurethane	310	416

In order to raise the fuel to its ignition temperature heat has to be transferred to, or generated within, the fuel. In those cases where this heat is generated totally within the fuel we speak of spontaneous combustion. This process can be chemical or biological in nature and the self-heating of haystacks due to bacteriological action is a well known example. Sometimes a rise in external temperature which is less than the ignition temperature can accelerate internal reactions so that heat builds up in the fuel to a stage where ignition occurs. Finally we have those situations where the heat energy necessary to promote combustion comes essentially from an external source referred to as the ignition source.

The ignition source must contain sufficient energy to raise the fuel to its ignition temperature and this energy must be transferred to the fuel rapidly enough for the ignition temperature to be achieved. If the source is a small flame or perhaps sparks or glowing embers, then we know the temperature will be high: usually within the range 600°C – 1000°C and possibly higher. The very fact that such a source appears luminous or glowing is an indication that the temperature is in this range. Temperatures such as this are well above the ignition temperature of most combustible materials as can be seen from Table 6 and under these conditions it is not only the temperature but also the heat energy available that is critical in determining ignition. Thus whilst they are at a high temperature, the sparks from a grinding wheel are unlikely to set alight a heavy wooden workbench. A flame from a match or a glowing cigarette end is at temperatures close to those attained in fully developed compartment fires. In most cases flame temperatures are of a similar order of magnitude irrespective of the size of the fire; it is the total heat energy available and how rapidly it is transferred to secondary materials that determines the processes of ignition and the spread of fire.

When an ignition source such as a flame is applied or is adjacent to a material, heat will be transferred to the surface of the material as illustrated in Figure 10.

The exposed surface of the material will become hotter. Some of the heat gained

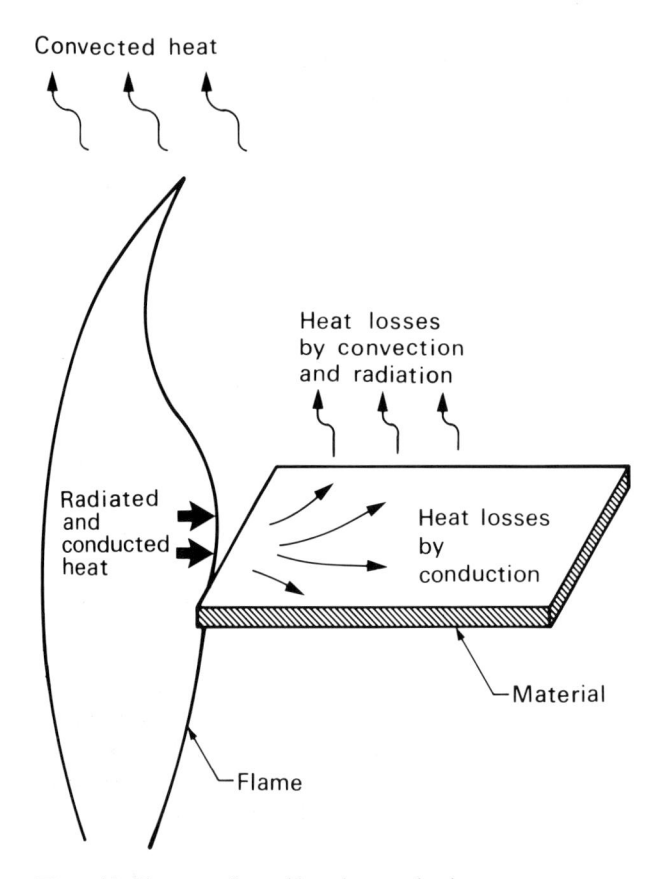

Figure 10 Heat transfer and heat loss mechanism

by the material will be lost from the heated surface by re-radiation as it becomes hotter, by conduction along the material and by convection. This balance in heat energy transfer can be as important as ignition temperature in determining the ease with which a material can be ignited. The heat capacity per unit volume of a material is a function of its density ρ and its specific heat c. The ease with which heat can be carried away from the heated area is measured by the thermal conductivity of the material k. It can be shown by theory that $k \times \rho \times c$ ($k\,\rho\,c$) is a measure of the time for a surface to heat up when exposed to a given heat source and hence is significant in determining the ignition characteristics of the material.

We are now in a position to explain what might appear to be anomalies in Table 6. If we use ignition temperature as a criterion it would appear from Table 6 that plastics should invariably show a markedly superior performance when compared with cellulosic materials such as wood. Many plastics, particularly foamed plastics, have very much lower values for $k\,\rho\,c$. When exposed to similar heat sources their surfaces will heat up much more rapidly and the ignition temperature is more readily achieved even though it may be higher.

These mechanisms apply equally to composite systems and if we are dealing with thin coatings such as paints the thermal properties of the substrate can have a marked effect on performance (Figure 11). In considering painted surfaces, mechanical properties such as bond and surface texture will also have a role to play and it is therefore important, particularly with fire retardant paints, to ensure that any performance evaluation is on the system intended for use.

Fire growth
Once ignition has been established the early stages of fire growth are subject to the same criteria discussed under the heading 'Ignition'. The early phases of growth must involve spread away from the source of ignition initially via the material first ignited. This spread is probably more important than the ignition itself as it is only when it spreads that the fire becomes dangerous. Fundamental to the mechanism of spread is the fact that the fire must be producing more heat than was necessary to promote the initial combustion reaction. The quantity of available heat from the material burning needs to be considered. The rate at which this heat can be released and fed back to the fuel is important in determining fire spread. We therefore have a system which may be represented approximately by the illustration in Figure 12.

It is evident from this simple model that if enough excess heat is generated and fed back to uninvolved fuel the growth of the fire will be an accelerating process. In the early stages of the fire, when it is small, interaction with the compartment will be negligible and the fire will behave as does a fire in the open. An important factor determining the initial rate of spread over and above those already discussed will be the proportion of the heat generated that is lost to the local environment. There will be many physical and geometrical parameters that control this, such as the shape and orientation of the fuel, the presence of edges and corners, the proximity of surfaces which may reflect back heat and, of course, draughts. Thus a cigarette smouldering on a cushion in the corner of a settee may well result in a fire which develops, whereas the same cigarette igniting a newspaper on a wooden table top may lead to a fire which extinguishes.

Once the fire becomes well established locally it will continue to grow as long as fuel and oxygen are available, subject to the limitations above. The heat referred to as lost heat in the description so far is still conserved within the system. Heat conducted into the fuel is warming the fuel so that less heat will be required to raise it to its ignition temperature. Heat lost by convection and radiation will be warming the compartment and adjacent combustible materials. Hot gases from the fire will rise to form a layer under the ceiling and as the walls warm up they will re-radiate heat. At this stage the fire can be referred to as a compartment fire and its progress will be influenced by the geometry and physical properties of the compartment. Ventilation can play a part in controlling the rate of growth at this stage and if restricted ventilation is holding back the rate of growth the situation can change dramatically if a door is opened or if windows break.

Soon flames reach the ceiling and are deflected horizontally. Often these ceiling flames are longer than one might expect. This is due to the slower oxygen entrainment in the hot gas layer, the unburnt gases from the fire having to spread

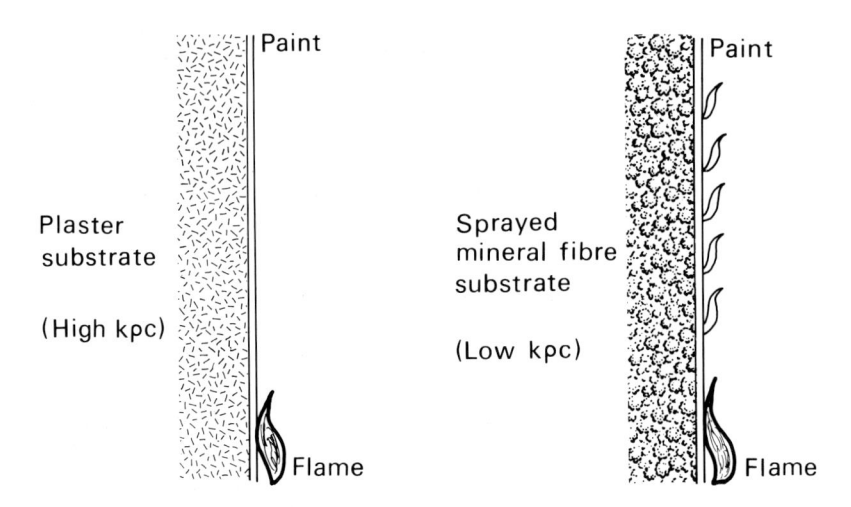

Figure 11 Influence of substrate

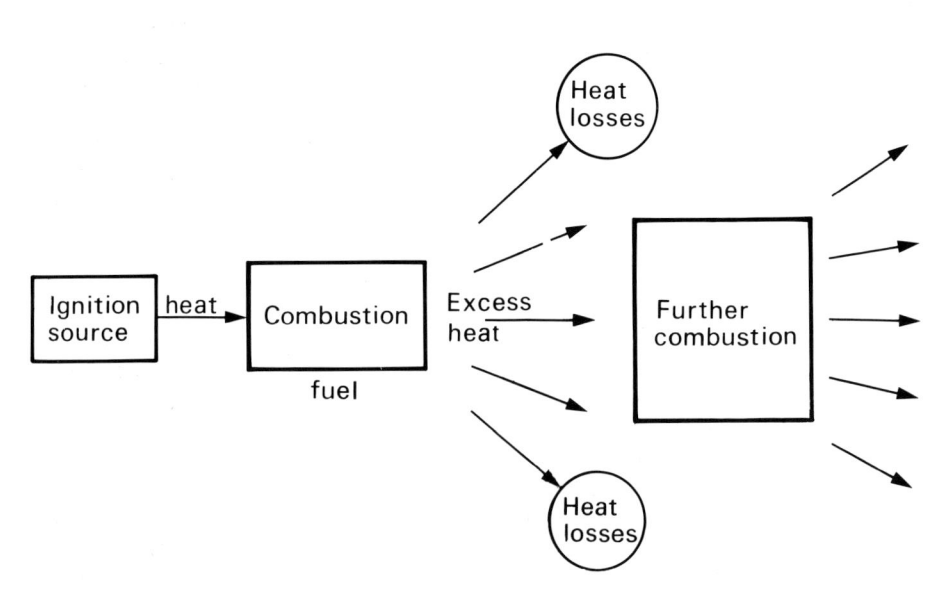

Figure 12 Initial fire growth

46

until they have mixed sufficiently for complete combustion. Combustible linings on the walls and ceilings play an important part in hastening the growth at this stage.

Whereas in the early stages of the fire convected heat was responsible for the general rise in room/compartment temperature, it is now radiation, particularly downwards radiation from the layer of flames and hot gases at the ceiling, that contributes to the increasing rate of fire growth. As the temperatures in the room increase, together with the downwards radiation, fuel remote from the site of ignition is being preheated. Eventually spontaneous ignition of contents distant from the flaming zone occurs. Generally this happens when temperatures in the hot gas layer reach 500°C – 600°C. Because of the sudden involvement of additional fuel the ceiling flames increase very rapidly, the hot gas layer deepens bringing the flames down closer to the fuel and all of the combustible materials in the room become involved. The atmosphere within the room may become more turbulent, encouraging mixing of unburnt gas with air, and the whole room appears to become full of flame. This transition from a growing to a fully developed fire is referred to as 'flashover'.

The developed fire

The post-flashover fire is often referred to as a fully developed fire. At this stage the fire almost invariably becomes ventilation controlled, the severity depending on the available air supply. Conditions within the fire compartment are such that the combustibles are being decomposed so quickly that it is not possible for sufficient air to enter the compartment to allow combustion to be completed. Therefore very hot combustible gases will spill out of the compartment, burning taking place as oxygen is encountered. Hence the familiar sight of long flames out of windows. More serious is the case when these hot gases find their way into further parts of the building through pathways such as doors and service ducts.

Usually it is openings such as windows which control the ventilation and hence the maximum rate of burning of the fire. As long as fuel is available the fire will continue to burn at this rate until it starts to run out of fuel and the decay phase is entered.

The relationship between the fire and fire tests

Regulations and/or supporting documents normally make use of standard fire tests to specify acceptable performance standards which materials or elements of construction would need to meet in defined circumstances. The United Kingdom and many other countries have their own national test standards and work is continuing within ISO to extend its range of standard fire tests.

In 1976 BSI, in referring to the limitations of fire tests[6], pointed out that 'Fire test methods are essentially attempts to assess the burning behaviour or performance of a material, product, structure or system under standardised and reproducible test conditions, which approximate to one or more stages of a real fire... No fire test, or combination of tests, can guarantee safety in a particular situation. They form only one of many factors which need to be taken into account in assessing fire safety'.

Tests which measure ease of ignition
We have seen that ignition temperature may be a poor guide to material performance because of other overriding factors. Many tests exist which examine the reaction of materials to small sources of ignition. It is important to ensure in the acceptance or adoption of a particular test that the test is related to the material application. A test of this nature is not intended to suggest that materials do not burn.

Bear in mind that the conditions of test and the ignition source are carefully defined in such a test and we are only obtaining an idea as to how the material will react to that ignition source or its equivalent. Such tests are often used for quality control purposes where, for example, they can be an indicator that flame retardants have been added. If a potential ignition source is known to be a probable event, for instance a discarded cigarette, then an appropriate ignition or ignitability test can be an indicator as to whether the materials under consideration can be used in that situation whilst still keeping the probability of a fire starting by that means acceptably low.

Tests concerned with the growth of the fire
Whereas under the previous heading we discussed the ease of ignition, we are now asking how readily materials burn once ignited. The process of flaming over surfaces is often used as a criterion for assessing the potential of a material to contribute to the growth of a fire. Usually the criterion is expressed as the duration of flaming, distance or rate of flame spread from the point of ignition. Flame spread away from the source will depend on the heat generated by the burning material itself and any additional heat available from what may be regarded as the source. Many materials when ignited continue to flame **only** when additional heat is available, ie the heat feedback from the burning material is not sufficient to allow a continuation of the burning process. In these situations, tests can measure the additional heat required to sustain flame propagation after ignition. Such tests usually subject samples to radiation gradients, and the point where flaming can no longer be sustained is a measure of the minimum supporting radiation necessary for flame propagation. Examples of such tests are BS 476 Part 7 (Method for classification of the surface spread of flame of products)[97] and BS 476 Part 3 (External fire exposure roof tests)[45].

As well as being concerned with such terms as flammability and flame spread it is also important to consider the overall contribution of the material to the fire. Some materials for instance may burn with small persistent flames whilst others may produce long luminous flames which irradiate neighbouring materials. Different materials make differing fuel contributions to the fire and the heat output of the material burning is an important factor in fire development. Tests with an open configuration such as the Surface spread of flame test of BS 476 Part 7 allow the escape of most of the heat generated by the burning material. An enclosed test such as the Fire propagation test of BS 476 Part 6[98] attempts to take into account the heat generated by the material. In this respect these two tests can be regarded as complementary as to the data that they provide about the fire hazard of materials.

As with ignitability tests it is important to interpret the results of flammability tests within the context of the specified test conditions. We must consider, for example, whether the test has been carried out with or without supporting heat radiation. Materials which are difficult to ignite or those which appear to burn with difficulty in normal environments can burn frighteningly quickly when the environment changes to the hostile one that accompanies a developing fire.

A test in the first instance is a method of ranking materials so that we can distinguish between the good, the less good and perhaps the downright bad. It is important not to lose sight of these objectives in a quest for simplicity, elegance and, from the laboratories' point of view, reproducibility.

It has been explained that fire is a complex and little understood phenomenon. We must not lose sight of the fact that generally we are using tests in a predictive way as a means of evaluating the fire hazard that may exist. Interpret and use test results with care, and above all ensure that the selection of a particular test is appropriate to the risk. Often this problem is solved for us by requirements which define the relationships between the test and its application. Even in these cases however care is needed in ensuring that the materials are tested in the right form. Paint films for instance will exhibit a performance which is intimately connected with the nature of the substrate. A paint film applied to steel or a plastered wall may not burn, whereas the same paint applied to sprayed asbestos or wood might contribute significantly to the spread of fire. Other materials may behave well in particular tests because of peculiarities in their properties. Materials may melt or decompose in the test apparatus; may swell up and affect the ventilation or the heat source, or may produce excessive quantities of smoke and carbon particles on burning.

Often it can be seen that the existing standard tests are insufficient to allow a satisfactory prediction of behaviour in fire. If this situation arises, fire researchers resort to what is termed *ad hoc* testing, where the real situation is modelled or, if necessary, a full-scale replica is produced incorporating all those features necessary for a realistic prediction of fire hazard.

Tests involving the developed fire
We have seen that the compartment fire does not spiral upwards in ever increasing severity but will eventually burn at a controlled rate usually dependent on the ventilation available. The fire has entered a period of steady state burning which can be referred to as the post-flashover stage. It is obvious that little can be done to control the fire within the compartment (passively) and therefore fire spread is now kept in check by trying to contain the fire. Since the total fire severity and hence the ease with which it can be controlled is a function of the amount of fuel burning, the effective containment of the fire to a limited area restricts the potential loss.

The philosophy of containment is embodied in a single test, that of Fire resistance, as measured in accordance with the appropriate national[99,100,101,102,103] or international standard. Fire resistance is a concept applicable to elements of the building structure. It is applied for instance to the walls, floors, columns and beams and is a measure of the period of time for which they would be able to withstand and contain a developed fire. Since the fire is fully developed within the compartment no account is taken of the contribution of materials to fire in these tests. Hence constructional materials such as wood, which burn, start off on an equal footing in comparison with non-combustible materials such as steel and concrete.

There is perhaps less diversity in tests for fire resistance than exists for tests used in other areas of fire hazard assessment. It would appear that at last a standard fire has been defined. This is in fact far from the truth and it is important to remember the following points. Firstly, in a fire in a building, the local fire severity will depend on the fuel available and the conditions under which it is burning. The standard fire resistance tests with a regularly defined fire condition provide a yardstick for assessing the relative performance of the structural elements. A thirty-minute fire-resisting door may well burn through in twenty minutes in a real fire; alternatively it may well resist the penetration of fire for three-quarters of an hour. What is certain is that its performance will be inferior to that of a sixty-minute fire-resisting door and that a non fire-resisting door would be likely to burn through much more quickly in the real fire. Secondly, it must be emphasised that fire resistance tests are dealing with the post-flashover fire. The time for a fire to develop is rather indeterminate, depending on the mode of ignition, the materials first ignited and a variety of local circumstances. A fire may take two minutes, two hours or two days to develop to flashover depending on whether the igniting source was say, a smouldering cigarette, or an arsonist at work. The only thing that we can be certain of is that the fire development will accelerate rapidly as flashover is approached. The standard time-temperature curves of the fire resistance tests are not intended to represent the growth of fires; they are more a reflection of the time constants of the test furnaces which are used for carrying out such tests.

The preceding paragraphs have discussed the applicability of tests and warned of their limitations. Armed with this information we may move on to a more detailed discussion of the performance of materials and currently available tests.

Chapter 5: Structural fire protection

In Chapter 1 it was explained that the object of fire precautions is to minimise fire hazard and that this could be achieved by precautions taken with the three aims of:

(i)   reducing the number of outbreaks of fire,
(ii) providing adequate facilities for the escape of the occupants, should an outbreak of fire occur, and
(iii) minimising the spread of fire both within the building and to nearby buildings.

The chance of controlling a fire following its discovery will depend on when the fire is discovered and on how quickly it is growing. Fire protection measures are therefore concerned with the control of the rate of growth of a fire and subsequently (if the fire becomes large) with its containment. These objectives can be achieved by paying attention to the building design, and by adopting methods and using materials which are designed to reduce the fire hazard.

Structural fire protection is therefore concerned with (iii) above and can influence fire growth, fire spread, structural stability and smoke movement. It is defined in BS 4422: Part 2[104] as 'those features in layout and/or construction which are intended to reduce the effects of a fire'.

Standard fire tests

The fire tests which are most frequently specified in respect of fire protection in the United Kingdom are those forming BS 476 which at present comprises 15 parts (including Part 10 which provides an introduction and general information). Use is also made of the quality control type tests specified in BS 2782 'Methods of testing plastics' to classify PVC and sheet plastics for use as light diffusers, ceiling panels, windows and roof lights.

A description of the BS 476 series, together with those BS 2782 tests used in connection with regulations, is given in Appendix 3. However, it should be noted that although BS 476: Part 3 was revised in 1975, the 1958 edition is still used in connection with building regulations.

The British Standards Institution, as part of its review of BS 476, allocated part numbers in advance so that the new standards were published in a coherent manner: Parts 11–19 applying to tests for products; 20–29 to the determination of fire resistance; and 30–32 to miscellaneous tests. Figure 13 shows the various tests related to the stages of an uncontrolled fire in a compartment.

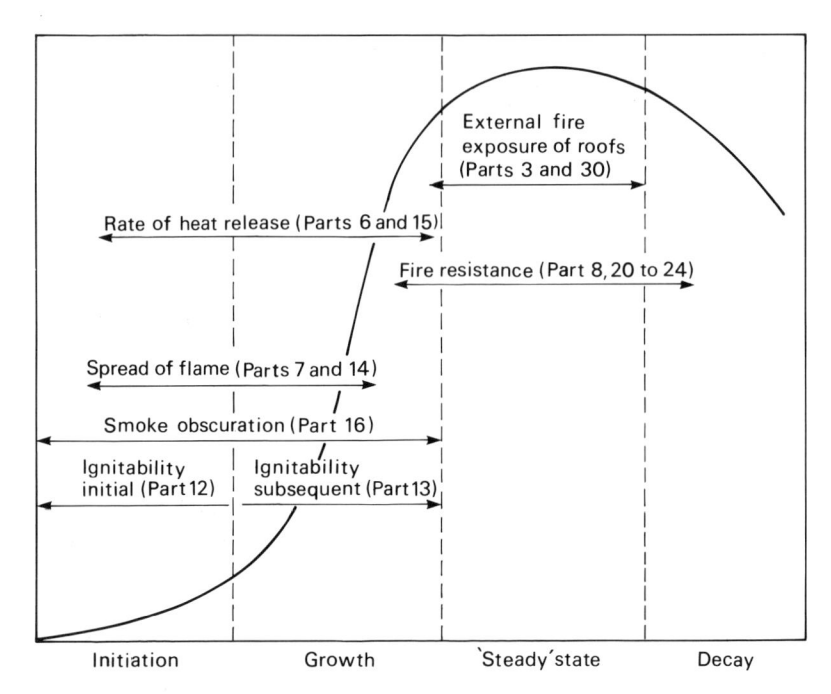

Figure 13 Fire phenomena and BS 476 Part numbers related to the stages of an uncontrolled fire in the compartment of origin

To assist in the removal of barriers to trade that might arise from the application of the different fire tests used throughout the European Community, a CEN* Committee was formed in 1987 to produce European fire test standards. CEN has been given a mandate by the European Community and the European Free Trade Association (EFTA) to prepare European Standards for fire resistance testing and for external fire roof tests. In due course these European Standards will replace the existing Parts of BS 476.

In 1976 the Department of the Environment (DOE) instituted a Supervisory Scheme for fire test laboratories undertaking routine testing of materials and products used in construction. This scheme was set up to provide an indication (particularly to local building control authorities) of the DOE's view of the competence of such laboratories to undertake fire tests specified in building regulations and certain other related tests. In 1984 the above scheme was incorporated within the National Testing Laboratory Accreditation Scheme (NATLAS)[†]. The independent laboratories currently undertaking BS 476 and/or BS 2782 tests include — British Gypsum Ltd (East Leake), the LPC Laboratories[‡] (Borehamwood), TRADA[§] (High Wycombe), Warrington Fire Research Centre (Warrington) and SGS Yarsley Ltd (Redhill). A directory of laboratories within the above scheme is published annually by the NAMAS Executive, National Physical Laboratory, Teddington, Middlesex TW11 0LW.

---

* Comité Européen de Normalisation
† now operated by the National Measurement Accreditation Service (NAMAS).
‡ formerly FIRTO.
§ Timber Research and Development Association.

The results of standard tests carried out by the Fire Research Station have been published[105,106,107,108] and the FPA (which assisted with most of these publications) has continued this work in respect of all UK laboratories undertaking sponsored testing under BS 476[109,110,111].

Fire growth

To reduce the probability of an outbreak and consequent growth in fire, where it is not possible to segregate all possible known ignition sources from combustible materials, the contents and room linings should be such that they are not easily ignitable and, even if ignition occurs, the ensuing fire will not develop rapidly. It should be noted that although materials can be treated to reduce their ease of ignition from a small source, such treatments may not affect their rate of burning once ignited.

Internal linings form large areas of continuous surfaces which may be heated by the burning contents of a room or compartment. When hot, these linings will heat (principally by radiation) materials not yet involved in the fire.

Some indications of the fire hazard of combustible lining materials are given by tests which measure the ease of ignition (BS 476: Part 5[112], now withdrawn), the rate of spread of flame over the surface under fire conditions (BS 476: Part 7[97]) and the rate at which materials will contribute heat to a fire (BS 476: Part 6[98]). However, the results obtained in such tests are heavily dependent on the exposure conditions and many other factors, and cannot be interpreted as describing fundamental properties of materials. Even materials classified as non-combustible under BS 476: Part 4[40] can affect the development of a fire. Materials with good thermal insulation properties contribute to a more rapid rise of temperature in the fire compartment; whereas materials of high heat capacity soak up heat from the fire and tend to slow its rate of growth (see Figure 14).

Wall lining	Flashover time
Dense non-combustible material (brick)	23 min 30s
Fibre insulating board with skim coat of plaster	12 min 00s
Hardboard with two coats flat oil paint	8 min 15s
Non-combustible insulating material	8 min 00s

Figure 14 Effect of wall lining on time to flashover in test rooms

Experience with real fires and the results of experimental work on linings in model rooms, combined with a knowledge of test ratings obtained by many types of materials, enable BS 476: Parts 6 and 7 to be used as a basis for controlling room linings in different situations.

Where linings are controlled, the general requirement is for Class 1* under BS 476: Part 7 but this is relaxed in carefully defined situations (for example, 'small' rooms are permitted to be Class 3). Where a higher degree of protection is required, for example in circulation spaces (which often constitute an escape route), Class 0 is specified.

Class 0 is a term introduced under the 1965 Building Regulations, as Class 1 covered too wide a range of performance for use in critical areas. The original definition included non-combustible materials (as now) and acceptable combustible linings†. In 1971 the definition for acceptable combustible materials was changed to a specified performance under BS 476: Part 6. In 1976 the requirement for the material to be also Class 1 was added as it had been found that some materials had achieved satisfactory fire propagation indices but were not Class 1. The definition under Approved Document B uses the term *material of limited combustibility* instead of non-combustible. This term (which embraces non-combustible materials) was introduced in 1985 in order to make reference to materials tested under BS 476: Part 11[113]. Therefore, as the definition has changed over the years and is not necessarily identical in all regulations/codes/guides, to avoid confusion it is preferable to quote Class 0 by reference to the date of the regulations or publication concerned.

Materials which do not reach the required performance under the above tests, and those which are unclassifiable because they fall away from the specimen holder before the test duration has elapsed, are permitted for use as lighting diffusers, ceiling panels, windows and rooflights. However, subject to use and location, limits are imposed on their flammability under BS 2782 (Methods of testing plastics) or BS 5438[114], the maximum areas acceptable, and the spacing between those areas.

Some examples of published performances under BS 476: Parts 6[107] and 7[108] are given in Tables 7 and 8.

Essentially fire spread over linings is a surface phenomenon. However, when considering a lining system for test, the sample must include any layer which

---

*See Appendix 3 for meaning of test ratings.
† (1965) (i) non-combustible throughout; or
       (ii) a Class 1 product comprising a surface ≤1/32" (0.8 mm) on a non-combustible base (substrate); or
       (iii) a combustible base having a non-combustible finish ≥1/8" (3.2 mm).
 (1971) (i) non-combustible throughout; or
       (ii) a combustible material having fire propagation indices I ≤12, i_1 ≤6.
 (1976) (i) non-combustible throughout; or
       (ii) a Class 1 product having fire propagation indices I ≤12, i_1 ≤6.

Table 7 Fire propagation indices for typical materials
(Values given are only indicative of likely performance and must not be taken as values for specific materials, thicknesses and substrates)

Material	I	i_1
Plasterboard	10	<6
Plasterboard with emulsion paint	9	4
Mineral fibre tile	8	<5
Woodwool slab (low density), untreated	10	5
GRP (flame retardant)	11	4
Wire reinforced PVC sheet	13	8
PVC sheet	17	6
Polystyrene rigid sheet	17	9
Softwood, with intumescent surface coating	18	<6
Fibre insulating board, as above	20	6
Expanded polystyrene containing flame-retardant additives	20	<15
GRP, untreated	32	22
Hardboard, untreated	36	<16
Plywood, untreated	41	19
Softwood, untreated	47	20
Acrylic sheet	51	<32
Fibre insulating board, untreated	66	41

Table 8 Some typical results of surface spread of flame tests
(Values given are only indicative of likely performance and must not be taken as values for specific materials, thicknesses and substrates)

Material	Class
Plasterboard	1
Plaster with 0.25 mm wallpaper	1
Woodwool slabs conforming to BS 1105: 1963	1
Timber, hardboard or fibre insulating board, with a good surface or impregnation treatment	1
GRP (flame retardant)	1
Phenol formaldehyde foam	1
Polyisocyanurate foam	1
Rigid uPVC sheet (when it can be classified)	1
Polycarbonate	1 – 2
PVC foam	1 – 3
GRP sheet (various types)	2 – 3
Hardwood, softwood or plywood, of density greater than 400 kg/m^3, untreated	3
Hardwood, softwood or plywood, with oil-based or polymer paint	3
Wood particle board, untreated or with oil-based or polymer paint	3
Hardboard, untreated or with wallpaper or with oil-based or polymer paint	3
Polymethyl methacrylate (PMMA) sheet	3 – 4
Polyurethane foam	4
Cellulosic fibre insulating board, untreated	4

(because of its thermal or combustible properties) is likely to affect the overall fire performance. Even when a non-combustible substrate is specified, its thermal properties can significantly affect the performance of a combustible finish, and the test result is directly applicable only to systems based on a substrate of similar thermal properties. If this substrate is thin, then the insulating board which is used to back specimens may affect the fire performance. Similarly, where the product under test consists exclusively of a thin or quickly destructible material, the backing board for test purposes may affect the result. Advice should therefore be sought before attempting to apply test data to systems incorporating other substrates.

No information on the effects of ageing is provided by these tests, which are normally carried out on new samples; neither can the performance of a multiple layer system be predicted from tests on individual components.

The choice of a lining system will frequently involve consideration of layered or laminated systems. One of the problems associated with delamination is the exposure of combustible substrates to the fire sooner than designed for. It is therefore important that the methods of fixing or bonding used on site will not permit premature delamination in the event of fire, resulting possibly in a markedly different response by an otherwise 'satisfactory' finish or veneer.

Untreated timber and its derivatives can generally achieve a Class 3 rating, but the use of flame retardant treatments can enable these materials to achieve Class 1 or Class 0. The types of treatment involve either surface application or impregnation, and their action varies. For example, the treatment may provide an impervious barrier over the surface of the material, or when heated it may react to form a layer over the material which inhibits combustion or insulates the surface. Information on flame retardant treatments for timber and its derivatives has been published by the FPA[115] and TRADA[116]. The selection of a suitable treatment may also depend upon practical considerations such as the type of surface to be treated, overpainting and durability. The durability of surface treatments should always be questioned, as many may be degraded by regular wetting, eg by rain or by condensation.

Certain plastics materials, because of their softening characteristics, cannot be assessed in the tests for fire performance of building materials. Thermoplastic materials soften and melt in a fire. Those which do so before they ignite do not contribute significantly to fire spread if they fall away from the flames, although falling hot materials may be a hazard to people beneath. Materials which ignite and burn before or whilst falling may encourage rapid fire spread. The performance depends on the types of plastics material, the thickness of the sheet and the way it is held in position.

Fire spread

Internal
The spread of the fire from the room of origin to other parts of the same building results from the same processes of growth as in the room of origin, ie by conduction, convection and radiation. The routes being firstly by passing through any existing gaps or openings in the surrounding construction (including doorways); secondly by burning through or opening up gaps within the construction; and thirdly by heat conducted through the construction — so igniting combustible materials in direct contact or in close proximity to the other side.

The fire and hot gases will then spread along any available routes such as corridors, stair and lift wells, service ducts and cavities.

When designing to prevent the spread of fire from the room of origin, various parts of a building may be required to be enclosed or separated by fire-resisting construction, the most notable technique being 'compartmentation'. A fire compartment is defined as 'a building or part of a building, comprising one or more rooms, spaces or storeys, constructed to prevent the spread of fire to or from another part of the same building, or an adjoining building'. Requirements for compartmentation are generally based on the occupancy and size of buildings.

It is therefore important to ensure that the separating elements of construction (walls, floors, supporting elements and doors) have the necessary fire resistance and that gaps or other openings in or between elements are properly fire-stopped. Attention must also be paid to the design of details involving ducts, pipes and services passing through fire-resisting walls and floors — particularly those elements forming the enclosures to fire compartments and to escape routes. A particular hazard exists in extensive cavities in horizontal or vertical constructions which, if fire penetrates them, act as 'chimneys or flues', conveying flames and hot gases over considerable distances; a situation which is aggravated when they are lined with combustible materials.

External
Fire spread from one part of a building to another part, or from one building to another, occurs by flames spreading up the external face of the building; by conduction of heat or passage of flames through a wall separating adjoining buildings; by radiation; by flying burning brands; or by flame projected across the space separating buildings.

The construction and behaviour of external walls (including cladding), party or separating walls (including the junction with the roof) and roof coverings, are therefore important in preventing the external spread of fire; as are the extent of unprotected window openings and the proximity of combustible contents and linings adjacent to windows.

Figure 15 shows the hazard which can be caused by radiation. In this case a fire in a furnishing store radiated sufficient heat to set fire to goods near windows in another building 22 m away.

The nearness of a building to the relevant boundary determines the possibility of its being at risk from a building on fire on an adjacent site. Thus any combustible cladding could be ignited by radiation unless sufficiently distant from any neighbouring risk. In addition, although the hazard of flame spread over the external face of a building depends largely on the building design and the method of construction of the cladding, the risk of ignition and subsequent flame spread from a localised fire source (such as flames emerging from a window) can be reduced by a suitable choice of cladding material. Whether or not restrictions are placed on the type of cladding or minimum performance ratings are specified, is governed by the probability of ignition by radiation (proximity to the relevant boundary).

Control in respect of external flame spread may also be imposed in certain situations. By specifying a cladding material in terms of its fire propagation indices under BS 476: Part 6, control is possible over the material's ease of ignition and rate of heat release — these two factors determining to a large extent the possibility of vertical flame spread.

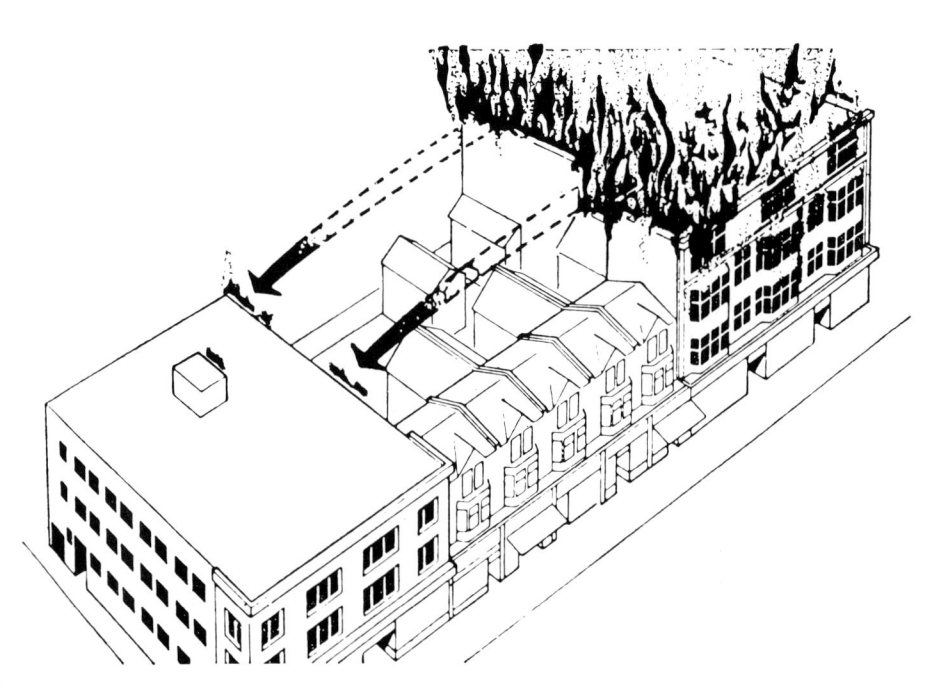

Figure 15 Radiation

British Standard BS 476: Part 3: 1958 is the method of test used at present for assessment of the possibility of ignition and extensive flame spread over the external surface of a roof caused by heat radiated by a fire in an adjacent building (or another part of the same building) and by flying burning brands and the risk that, once the roof has been ignited on the outside, the fire will penetrate to the inside of the building (Figure 16). However, this test cannot be used for some forms of roof construction. For example, some types of plastics materials may not ignite but will fall away prematurely and so be unable to be classified.

In general one must consider the fire performance of a roof as a whole, including any internal lining. Not only is the degree of combustibility of the roofing materials relevant, but also such factors as their softening or melting temperatures; the fixing and jointing methods used; the thermal properties of any insulation; and the supporting structure.

Results of roof tests carried out at the Fire Research Station were last published in 1970[105]. Other data exist in tables of deemed-to-satisfy forms of construction.

Acceptable forms of roof coverings for the purposes of regulatory control are given by reference to their proximity to the boundary and to the size and purpose (occupancy) of the building concerned. Although generally specifying minimum roof-to-boundary distances by reference to designated roof coverings, provision is also made for the use of named materials (eg thatch and wood shingles) and for those which cannot be designated (eg certain plastics materials).

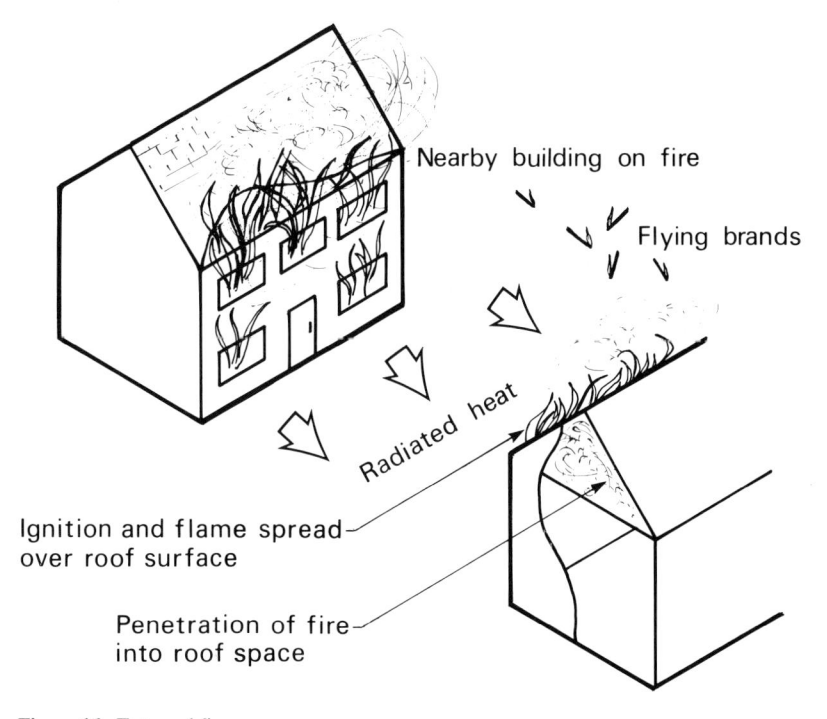

Figure 16 External fire exposure

Structural stability

The need for maintenance of the stability of a building on fire will depend upon various factors such as:
(a) Is the building expected to survive a fire or be demolished?
(b) To what extent is structural stability of the building required for:
 (i) means of escape,
 (ii) the effectiveness of separating elements of construction, or
 (iii) firefighting?

For any building, therefore, the parameters for determining the need for structural stability will be determined by the owners'/insurers' needs and the location, size, height and type of occupancy of the particular building.

Structural stability in fire is achieved by the provision of adequate fire-resisting construction and the correct detailing of junctions between different elements.

Fire resistance of elements of building construction

The property of fire resistance can be regarded as the ability of an element of building construction to fulfil its designed function in the event of a fire. This function may be to contain a fire (as with a non-loadbearing wall); to support the design load (as with a beam or column); or both (eg a floor). It should be noted that fire resistance relates to an element of building construction and is **not** a property of a material; however, the properties of a material will affect the performance of any element of building construction of which it forms part. Desirable properties which are relevant include — dimensional stability; adhesion; cohesion; the ability to resist thermal shock and cracking; and low thermal conductivity. However, the way in which a material is used, the design of the element of building construction and the manner of fixings, are all important.

Regulations and supporting documents specify the minimum periods of fire resistance necessary in defined situations — generally according to the occupancy of the building and to its height or cubic capacity. Although BS 476: Parts 20 to 23[99,100,101,102] indicate which criteria are appropriate for the different types of element, situations are also identified where the full criteria are not required; where the duration for each criterion is not identical; and where the element is not expected to perform equally from both sides (see Tables 9 and 10).

Information on the design of fire-resisting elements of building construction can be obtained by reference to:
(a) specifications tested, or assessed, under the relevant part of BS 476,
(b) appropriate British Standard Specifications or Codes of practice, or
(c) deemed-to-satisfy specifications.

Tables of notional periods of fire resistance for walls, concrete structures, structural steelwork and timber floors are included in a BRE Report[117].

It should be stressed that consideration may also need to be given to the probable

Table 9 Examples of different 'half-hour' periods of fire resistance

Half-hour element	Expected minimum performance (minutes)		
	Loadbearing capacity* (Stability)	Integrity	Insulation
External wall	30	30	30
	30	30	15[†]
Floor	30	30	30
	30	15	15[‡]
Glass	—	30	Nil[§]
Door	—	30	Nil[¶]
	—	20	Nil[¶]

* Not applicable to non-loadbearing elements.
† If 1 m or more from relevant boundary (see Table 10).
‡ Applicable only to two-storey dwelling houses.
§ Ordinary fire-resisting glass is unable to attain practical periods of insulation.
¶ Doors are not usually required to afford insulation.

Table 10 Provisions as to method of test and minimum period of fire resistance for walls (Approved Document B, 1992 Edition)

Part of Building	Minimum provisions when tested to the relevant Part of BS 476 (Minutes)			Method of Exposure
	Loadbearing capacity* (Stability)	Integrity	Insulation	
Any part of an external wall 1 m or more from the relevant boundary	●	●	15	From inside
Any part of an external wall less than 1 m from any point on relevant boundary	●	●	●	From each side separately
Compartment wall	●	●	●	
Compartment wall separating buildings	● (min 60)	● (min 60)	● (min 60)	
Wall separating an attached or integral garage from a dwelling house	30	30	30	From garage side

● denotes period of fire resistance specified in Table A2 of the Approved Document.
* Not applicable to non-loadbearing elements.

effects of mechanical damage, weathering, thermal movement, etc when selecting the construction, or mode of protection, for an element of structure. Equally, the objectives of regulations and codes of practice (and of designers) to achieve buildings which are fire-resisting will be lost by bad workmanship. Good quality control is important in the use of all materials and composites — particularly with those fire-protecting materials which can be applied by unskilled labour.

Concrete, steel and timber structures
The fire resistance of a reinforced concrete structure depends, to a large extent, upon (a) the overall thickness of the section (in order to keep heat transfer through the member within acceptable limits); and (b) the average concrete cover to the reinforcement (in order to keep the temperature of the reinforcement below critical values). The tendency of concrete to spall, or break up, in a fire can lead to loss of the insulating cover to the steel and reduction in overall thickness of the member. In some constructions supplementary reinforcement is necessary to reduce these effects. The overall thickness and cover is determined by the properties of the aggregate used. For example, lightweight aggregate formed from expanded pulverised fuel ash has a low thermal conductivity and expansion, and is resistant to spalling — this will enable thickness and cover reduction to be made without lowering the fire resistance. Additional information on the fire resistance of concrete is given elsewhere[118].

Steel has a high thermal expansion which can lead to thermal movement in other parts of the structure, which in turn may cause loss of integrity and stability. Unprotected structural steel cannot normally achieve practical periods of fire resistance by virtue of the relatively low temperature of 500°C – 550°C (for mild steel) at which it loses 50% of its strength. Failure will occur if the safety factor with respect to ultimate load is 2 (a normal figure in engineering practice). To satisfy the criterion of loadbearing capacity therefore, a loaded steel section generally needs to be protected to ensure that this temperature is not attained during the fire resistance period required — due consideration being given to the way in which continuity, restraint, etc may affect this value.

The performance of structural steel depends on the mass of the steel and the surface area exposed to heating. In the past, steel section sizes adopted for fire tests were identical and thus deemed-to-satisfy tables referred to sections of a minimum mass. In order to provide design information for the protection of a range of steel section sizes, the Fire Research Station in conjunction with industry developed a programme of fire resistance tests designed to provide a computational method of assessment for the thickness of protection needed. Methods of affording protection to steel beams and columns by passive encasement systems and by the use of intumescent coatings have been published by the ASFPCM (Association of Structural Fire Protection Contractors and Manufacturers, now the Association of Specialist Fire Protection Contractors and Manufacturers)[119]. Other less conventional methods, including water cooling of hollow steel sections and flame shields, are reviewed elsewhere[120,121].

Timber has a low thermal expansion which minimises the possibility of charred material or other protective layers becoming displaced. It also has a low thermal conductivity which means that timber immediately below the charred layer retains its strength. As a general rule it may be assumed that timber will char at a constant rate when subjected to the standard heating conditions in a test furnace. The rates of reduction in size of structural timber given in Table 11 apply for each face exposed. Different rates apply when all faces are exposed. It should be noted that these rates cannot usually be improved by the use of flame retardant treatments. Methods of calculating fire resistance of timber members are given in BS 5268: Part 4: Section 4.1[122].

Table 11 Charring rates of timber*

	Species	Depth of charring in 30 mins	60 mins
(a)	All structural species listed in appendix A of BS 5268: Part 2: 1989 except those noted in items (b) and (c)	20 mm	40 mm
(b)	Western red cedar	25 mm	50 mm
(c)	Hardwoods having a nominal density not less than 650 kg/m^3 at 18% moisture content	15 mm	30 mm

*From BS 5268: Part 4: Section 4.1

The design of concrete, steel, masonry and timber elements is discussed by Malhotra[123].

Glazed elements
Where annealed glass is incorporated in the design of an element of construction it must be accepted that there will be local high heat transmission and radiation which cannot be reduced for all practical purposes by increasing the thickness of glass. In this respect annealed flat glass is not able to satisfy the insulation criterion for more than a few minutes and some glass blocks can do so for no more than about 15 minutes.

Normal glass may shatter readily under fire conditions and therefore the types of glass traditionally used in fire-resisting constructions are those specified in CP 153: Part 4[124] (now withdrawn), namely (a) wired glass and (b) unwired glass forming copper-lights and small vision panels in doors. However, because such glass cannot satisfy 'insulation' it cannot meet the performance required for separating walls, compartment walls and walls enclosing protected shafts, and both Approved Document B and BS 5588 restrict its use where radiation would prejudice persons attempting to escape. Recently some unwired types of glass have been developed which have achieved satisfactory performances in terms of integrity and some have also satisfied insulation for 30 minutes or more. Guidance on the fire performance of glazed elements is given in PD 6512: Part 3[125].

Walls
Depending upon its situation and function within a building, a wall may be expected to fulfil different requirements in the event of fire. Most fire-resisting walls used to separate buildings, enclose compartments and contain fire, will be required to provide a barrier to the passage of fire from one side or the other and must therefore be able to satisfy each of the relevant criteria (integrity, insulation and — if the wall is loadbearing — loadbearing capacity) from either side for the prescribed period. Other situations arise where fire resistance is not required from both sides and where the construction may have to satisfy the criteria to different extents. Examples of different wall locations/requirements were given in Table 10.

In the case of external walls, the nearness of a building to the 'relevant' (facing) boundary determines the probability of its being a danger to other buildings on adjoining sites (if it is on fire) or of it being at risk from a neighbouring building on fire. Requirements made in connection with building regulations therefore specify different performances for external walls depending upon their distance from the relevant boundary. Where the walls are permitted to provide fire resistance only from the inside, loadbearing capacity and integrity are required to be satisfied for the full period; whereas insulation is required for only 15 minutes. This means that satisfactory constructions will be very different from those required to maintain insulation for the full period and where fire resistance is required from either side.

Loadbearing walls occasionally form part of the structural frame of a building without performing a separating function. In this event the construction would be judged only by the criterion of loadbearing capacity. Such a wall, while it may have to withstand the effect of fire from both sides at once, is difficult to test in existing designs of furnace which apply heat to only one side. Most constructions which are satisfactory when tested from each side separately are not necessarily adequate when heated from both sides at the same time.

In the case of framed walls, the contribution to integrity and insulation of the facings on each side of a symmetrical framed construction cannot be assumed to be identical. Once the exposed facing is penetrated, the unexposed side will be attacked on its inside face and will generally fail in a shorter time. This being so, additive methods of computing fire resistance require careful consideration: the nature and thickness of facings; stud size and spacing; type, thickness/density and method of fixing cavity insulation; and loading conditions; are all important. Possible areas of weakness in such walls are — joints and junctions, method and type of fixings, charring of combustible framework and the expansion of metal studs. Methods of calculating fire resistance of timber stud walls and joisted floor constructions form the subject of BS 5268: Section 4.2[126].

Floors
The method of test for floors is by heating from the underside. In the case of floor constructions utilising a ceiling membrane to contribute to the overall fire

resistance, such a membrane may be performing a number of functions:

(a) protection to structural members (beams or joists) only,
(b) prevention of fire penetration through the floor, or
(c) both (a) and (b).

Once the ceiling of a timber suspended floor has been penetrated, the supporting beams and/or joists and the underside of the floorboarding become exposed to fire (Figure 17). Although the fire resistance of an element of construction applies to the complete assembly, the stability of any individual member supporting the floor must be considered. Plain edged (or badly fitting tongued and grooved) boarding when used as flooring contributes little to the fire integrity of a floor and the time to compliance with integrity and insulation is effectively the collapse time of the ceiling — whereas close fitting tongued and grooved boarded floors can contribute to integrity and insulation for up to 10 minutes or so. Where all joints are protected, greater contribution is afforded.

Where consideration is being given to the use of a fire-protecting suspended ceiling as a means of contributing to the fire resistance of a floor, it should be noted that many of these ceilings may have been tested only in respect of the protection which they afford to steel beams, and therefore the criteria by which they have been judged may not be appropriate for floors.

Roofs
On fairly rare occasions a roof may be required to be able to contain a fire. For example, where it forms part of an external escape route or is required to afford protection against fire spreading to higher storeys of the same development from a podium or other lower structure. The roof must then be constructed to achieve the necessary period of fire resistance (tested as for floors, ie from beneath). It will also be necessary for any supporting structure to be fire-resisting.

In some situations, although the fire resistance of the complete roof is not necessary (ie to contain a fire), it may still be necessary to protect the supporting structure, eg where the collapse of the roof structure might otherwise endanger the stability of other fire-resisting elements. In connection with the design of certain portal framed structures where it is not desired to protect the rafters, guidance has been published by the Steel Construction Institute[127].

Guidance on fire precautions in connection with thatched roofs is available from the Rural Development Commission[128].

Fire doors

Where different parts of a building need to be separated from each other by fire-resisting construction (eg protected escape routes and compartments) the doors in any openings in such enclosures have precise functions to fulfil in case of fire. First, the door assembly should prevent the passage of excessive amounts of combustion products (including smoke) which could interfere with the safe use of

Figure 17 Underside of timber joisted floor construction after test, with ceiling protection removed

escape routes; second, it should maintain the effectiveness as a fire barrier of the wall or partition in which it is located. Every fire door is therefore required to act as a barrier to the passage of smoke and fire to varying degrees. An explanation of the logic of the provision of fire doors in relation to precautions against fire was given in PD 6512: Part 1[129].

The method of test for judging the fire resistance of door and shutter assemblies is that given in BS 476: Part 22, and the performance of door and shutter assemblies as barriers against the passage of smoke forms the subject of BS 476: Part 31. Testing for ambient temperature conditions is covered by Section 31.1[130] and methods are under development which describe procedures for evaluating performance at medium and high temperatures (Sections 31.2 and 31.3).

As regulations and codes of practice have not generally required compliance with the criterion of insulation, it has been practice to refer to doors by a two figure notation, eg 30/20, 30/30, 60/60 (in each case the first number being the number of minutes for 'stability' and the second for 'integrity'). However, reference is made only to integrity in Approved Document B, and in BS 5588 under which doors are designated FD20, FD30 or FD60 with a suffix 'S' when required also to retard the passage of smoke at ambient temperature conditions. This system is also used in Approved Document B and the Technical Standards supporting the revised building regulations for Scotland.

The term 'fire-check' door stemmed from the title of BS 459: Part 3[131] (first published in 1946, but now withdrawn) in which constructional specifications were given for two types of flush door. Such doors, although being able to provide effective barriers to the passage of fire for half an hour and for one hour, did not meet completely the criteria laid down in BS 476: 1932[37]. BS 459:Part 3 also provided that any door, however constructed, which gave a similar performance to the doors specified (generally assumed to be 30 minutes' stability and 20 minutes' integrity in the case of the half-hour door, and 60 minutes' stability and 45 minutes' integrity in the case of the one-hour door) could be termed a 'fire-check' door. Other terms which have been used to describe fire doors are 'Type' as provided in CP 3: Chapter IV: Parts 1–3[27,28,29], and 'Class' in the London Building (Constructional) By-laws.

For many years timber and timber-based doors very seldom achieved performances better than half an hour (many in fact were only able to qualify as half-hour 'fire-check' doors). Early designs for timber door assemblies generally used the frame profiles specified in BS 459: Part 3 but the leaves were of solid construction. However, in recent years the industry has successfully developed door assemblies of different core compositions, thickness, weight and frame profiles. Some of these have achieved performances in excess of one and a half hours, even with two leaves opening through 180°. Therefore it is no longer appropriate to presume that timber-based doors should be 45 mm thick (for half-hour) or 54 mm (one-hour); that a 25 mm wide stop/rebate is essential; or that double leaf assemblies require rebated meeting stiles. Neither is a 25 mm rebate or stop necessary for the purposes of smoke control, as the prime factor is the fit of the door in the frame, ie the smallest gap between either the door edge and frame, or the face of the door and the rebate or stop.

In cases where doorsets are not used, it is essential for the purchaser/specifier to obtain guidance on the type of frame needed to give the required performance for the particular door(s) selected. Guidance should also be sought regarding the appropriate specification for hinges and other door furniture. A number of doors are now sold with a colour-coded plug fitted in the door edge which identifies the potential fire resistance of the door leaf and whether or not site-applied intumescent protection is needed in the frame. Similar systems of colours are used by TRADA and the British Woodworking Federation (BWF), except that TRADA use a tree motive for the centre colour.

British Standard Code of practice BS 8214[132] gives guidance on the specification, design, installation and maintenance of fire doors with non-metallic leaves.

Figure 18
'Any attempt to prevent a fire door closing....'

It will be appreciated that any fire door will only fulfil its function of containing the spread of smoke and/or fire if it is fully closed (or is able to close automatically) at the time of a fire. Any attempt to prevent a fire door closing by the use of wedges, hooks, fire extinguishers, furniture, etc (Figure 18) or by disengaging or otherwise tampering with the closing mechanism, may have disastrous effects in the event of a fire.

Unless a fire door also performs functions required by the building users, eg for amenity or security, it will be considered an unreasonable nuisance and is likely to be held open. In these circumstances, and with those doors fitted with closers which users (particularly people with disabilities) find great difficulty in opening, various types of system are available which can reduce the probability of abuse. A British Standard specification has been produced for certain automatic release mechanisms[133].

Smoke movement

Smoke is defined in BS 4422: Part 1[3] as a 'visible suspension in atmosphere of solid and/or liquid particles resulting from combustion or pyrolysis'.

A fire mixes with the surrounding air, heats it, reduces its oxygen content and contaminates it with combustion products — gaseous, liquid and solid — which can be irritant, noxious and toxic. The most predominant toxic product is generally carbon monoxide. The concentration of smoke particles depends on the materials burning and on the conditions of combustion (particularly whether combustion is flaming or smouldering) as well as on the temperature and the supply of air.

A fire, even a small one, produces vast quantities of smoke and hot gases which can penetrate far beyond the flames and, although they eventually cool, they still constitute a hazard. A very high degree of dilution with fresh air is required to reduce the smoke from an average fire to a concentration which would be acceptable in an escape route.

The hot smoky gases from the fire rise upwards until they reach an obstruction (usually the soffit of a ceiling or roof) where they spread laterally and initially form a stratified layer. This layer deepens and may eventually fill the entire building. For example, in a couple of minutes, the smoke produced by an upholstered piece of furniture situated at ground level may possibly be enough to prevent escape from the upper floor of a house.

There are two main factors that determine the movement of smoke arising from a fire in a building (Figure 19). These are:

(a)   the mobility of smoke that results from its temperature and lower density;
(b)   the normal air movement, which may have nothing to do with the fire, that can carry cold smoke (sometimes slowly, sometimes quickly) to all parts of the building.

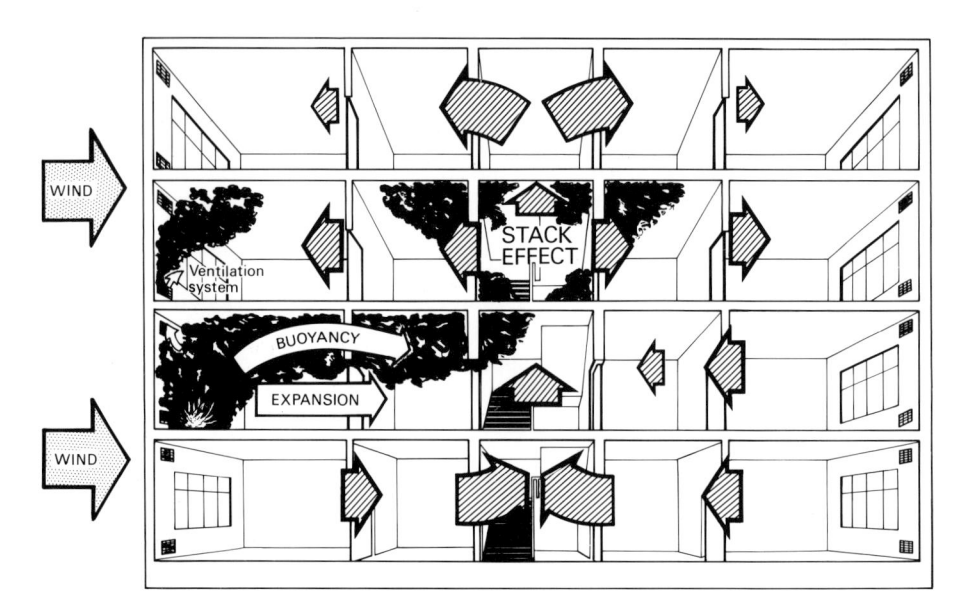

Figure 19 The forces driving smoke through a building

The air movement is itself controlled by:

(i) the stack effect*;
(ii) the wind (all buildings have some air leaks, and wind action contributes to air movement through them);
(iii) any mechanical air-handling system installed in the building.

Smoke will exploit any cracks, openings, cavities and ducts or shafts, etc. The factors involving smoke production and movement are therefore the materials involved and their rate of burning, the existence of any routes by which the smoke can travel and pressure differences.

Measures which can be taken to reduce the hazards presented by smoke include the selection of materials to reduce the chance of ignition and/or excessive smoke production, and by incorporating features into the building design to control its spread. Unwanted paths which can be exploited by smoke and hot gases should be 'fire-stopped'. Smoke control doors may be provided in corridors, etc, possibly in association with pressurization. In some cases the provision of venting or forced extraction can be beneficial.

---

*Pressure differential caused by the air inside the building being at a temperature different from that of air outside the building, which (when there are openings top and bottom) will promote natural airflow through the building — upwards when the building is warmer than the outside air; downwards when it is cooler.

Roof venting

Roof venting can be used to reduce the damage by heat and smoke and to improve conditions for escape and/or firefighting. The principles for venting smoke and reducing the damage by heat have some points in common, but in general very much larger vent areas are required for the latter.

In general, early but localised failure of the roof membrane (as opposed to the structural framework) can have considerable beneficial effects from the point of view of escape for the occupants and for firefighting conditions in the fire compartment. This is because venting will release some of the hot combustion products. Given a modest fire, if there is an adequate area of openings in the roof and an appropriate area of inlets for fresh air at a low level in the building, the combustion products can be restricted to a layer beneath the roof, leaving the atmosphere relatively cool and clear at floor level (Figure 20). The effectiveness of venting is greatly increased if the area beneath the roof is subdivided into 'smoke reservoirs' either by the design of the roof (Figure 20) or by 'screens' provided for the purpose (Figure 21).

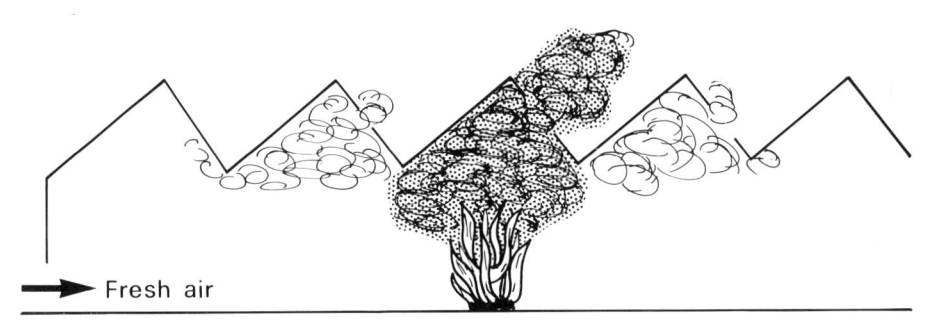

Figure 20 Roof venting

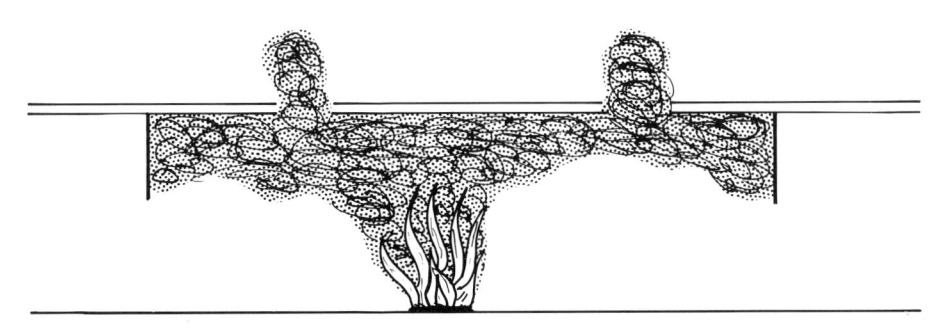

Figure 21 Screens beneath a flat roof

Venting for clearing smoke to facilitate escape — eg in enclosed shopping complexes — and possibly for firefighting, should be effective at as early a stage in the fire as possible. Such venting is generally best achieved by the use of roof ventilators, preferably operated by smoke detectors, although in some circumstances mechanical ventilation by fans may be more appropriate. Where mechanical ventilation is used there are many factors to be taken into account to ensure effectiveness; these are discussed elsewhere[134,135].

Fire venting may also be achieved by using rooflights of a thermoplastic material provided it falls away at an early stage of a fire before actually igniting. Rigid PVC and polycarbonate are two materials which have been considered suitable for this purpose. However, these rooflights are likely to collapse at a much later stage in the fire than that at which mechanical vents are set to open, but they have the advantage that larger vent areas can be provided economically. Because significantly larger amounts of flammable gases will be released, the length of the flame under the roof will be reduced — localising the effects of the fire (compare Figure 22a with 22b). This can have a beneficial effect on fire spread in some types of occupancy but where the fire can spread rapidly, unaffected by the roof — eg in a warehouse with high piled stock such as combustible cartons — the benefits are lost. Also in some cases it may be necessary to consider the effect of venting on the spread of fire to other parts of the same building or to other buildings nearby.

Materials which are likely to fall before ignition could nevertheless be hot enough to cause a hazard to firefighters. It is safer to have the entire roof membrane of a low melting point material because this will permit localised collapse before ignition over the site of the fire (Figure 22c) instead of at some distance away from it. This approach will also almost eliminate the role of the roof in fire spread provided that there is an adequate area of inlets for replacement air at low level.

The use of PVC rooflights for venting fires in single-storey buildings forms the subject of Fire Research Technical Paper No 14[136].

(a) No venting

(b) Venting by plastic rooflights

(c) Low melting point roof membrane

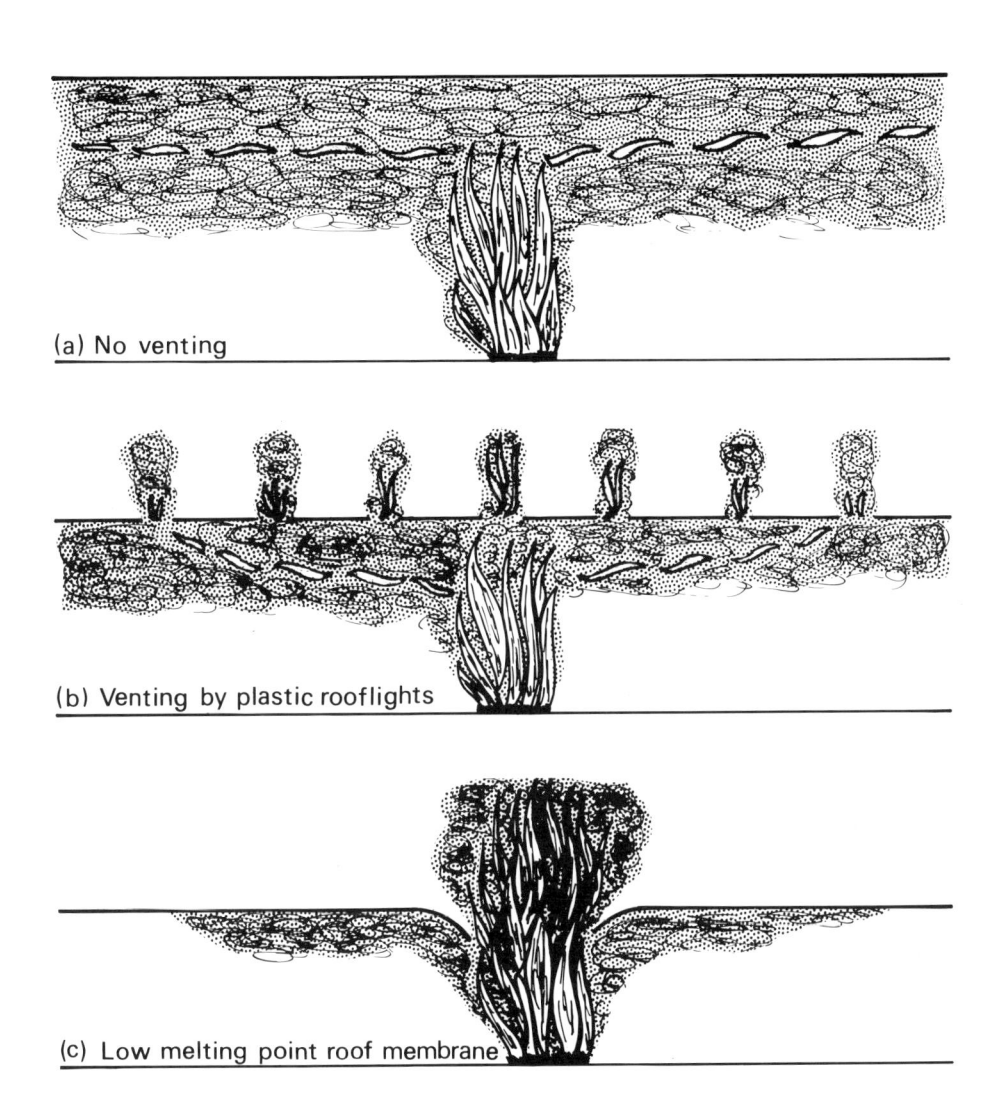

Figure 22 Effect of venting on flames beneath a ceiling

Chapter 6: Means of escape in case of fire

The Fire Grading Committee in Part III of their report[22] stated that 'fundamentally means of escape is but one aspect of the wider problem of circulation within a building. In view of the close relationship between normal circulation and means of escape it is evident that the latter cannot be divorced from general planning considerations and treated by a series of hard and fast rules. Accordingly it is only possible to establish certain principles which take into account the special conditions arising in a fire'.

For the purposes of their report, the Committee defined means of escape as 'structural means which form an integral part of the building, whereby persons can escape by their own unaided efforts'. This definition differs only marginally from that given until recently in BS 5588, ie 'structural means whereby a safe route is provided for persons to travel from any point in a building to a place of safety by their own unaided efforts'. Hence, reliance for fire safety on manipulative apparatus for means of escape, or on external rescue by the fire service using mobile ladders, is not considered satisfactory.

However, with the development of the code of practice for means of escape for disabled people (BS 5588: Part 8[137]) and the recognition that the successful emergency evacuation of a building using the structural means of escape provided requires comprehensive management procedures, whether the occupants are disabled or not, reference to 'by their own unaided efforts' has been deleted. Also lifts with special safeguards are now accepted for the evacuation of persons with a mobility handicap.

For the meaning of other terms used in connection with means of escape in case of fire, BS 4422: Part 6[138] applies except where the particular document or code of practice otherwise provides. The most important terms used are:

final exit — the termination of an escape route from a building giving direct access to a street, passageway, walkway or open space, and sited to ensure the rapid dispersal of persons from the vicinity of a building so that they are no longer in danger from fire and/or smoke.

storey exit — a final exit, or a doorway giving direct access to a protected stairway, firefighting lobby, or external escape route.

NOTE Under Section 5 of the Fire Precautions Act 1971 (as amended by the Fire Safety and Safety of Places of Sport Act 1987) 'escape' is defined such that the fire authority can take into account matters affecting the means of escape where it extends beyond the premises to a place of safety, ie a place in which a person is no longer in danger from fire.

travel distance — the actual distance to be travelled by a person from any point within the floor area to the nearest storey exit, having regard to the layout of walls, partitions and fittings.

Requirements

Requirements for means of escape in case of fire from new buildings in England and Wales, and Northern Ireland, can be met by following the guidance given in the documents supporting the respective building regulations[52,139] or by using other relevant codes of practice, eg BS 5588. In Scotland, the mandatory Technical Standards[54] should be used.

British Standard Code of practice BS 5588 gives guidance on how to design effective means of escape in case of fire from new —

- Residential buildings (Part 1)[30]
- Shops (Part 2)[31]
- Offices (Part 3)[32]
- Places of assembly (Part 6)[140]
- Shopping complexes (Part 10)[141]

Other documents dealing with means of escape from new buildings have been published by the Department of Education and Science[142], the Department of Health[143,144] and the Scottish Office[145].

Guidance in connection with existing premises has been published for hotels and boarding houses[67]; factories, offices, shops and railway premises[68,80]; places of entertainment[146]; hospitals[147]; residential care premises*[148]; and for houses in multiple occupation[92,94].

Considerations

Current requirements, whether controlled by specific rules or guidance, or by reference to codes of practice, have a common aim to secure safety from fire by:

(a) planning sufficient escape routes and adequately protecting them against the effects of fire from all parts of the building,
(b) planning to prevent the spread of fire and, in particular, to control or remove smoke and other combustion products in escape routes, and
(c) ensuring, by design, construction, finish and equipment, that the building structure will provide an adequately fire-safe background for the provision of means of escape.

However, in designing for means of escape, it is generally conceded that it is difficult to make comprehensive recommendations capable of covering every possible risk, and an intelligent appreciation of the principles and application of

---

* eg children's homes, community homes, homes for the elderly, and homes for the mentally ill and the mentally and physically handicapped.

the recommendations contained in any guidance document is essential. For example, BS 5588 explains that the only sound basis for designing means of escape from fire is to attempt to identify the positions of all possible sources of outbreak of fire and to predict the courses that might be followed by the fire as it develops, or, more particularly, the routes that smoke and hot gases are likely to take, including concealed spaces. Only against this background is it possible to design and protect escape routes with some confidence that they will be safe.

After a fire has broken out, there may well be a very short time during which the actions necessary for ensuring the safety of occupants can be carried out. This time will only be sufficient if all contributing factors (such as the design of the building, the functioning of alarm and detection systems, escape procedures, etc) are planned and managed so as to be effective when fire occurs.

A fire occurring anywhere within a building must be regarded as offering an immediate risk to all occupants within the compartment of origin — even though in the initial stages of fire development it might seem that people are well removed from imminent danger. It must be realised too that there may also be a risk to people in other parts of the building, as the effects of the fire will usually not be confined to the space in which it originated.

These considerations are particularly important when dealing with large numbers of people — some of whom may be unfamiliar with their surroundings and who may also vary widely in age and degree of mobility.

The following points are therefore relevant in the planning and protection of escape routes.

Type of risk and number of occupants
The type of risk (or occupancy) of the building under consideration will be determined by, to a great extent, the probable number and distribution of the occupants, their physical condition, and the way in which they may be expected to behave. For instance, according to the type of building, the occupants may or may not be familiar with the premises. People may be sleeping in the premises or sleeping above a portion of the building which is occupied only during business hours. The building may be visited by members of the public (possibly in large numbers) or people of a particular age group or degree of mobility may predominate. For example, in hospitals it may not be practicable to consider evacuation by way of stairs . Therefore the design for means of escape will need to provide for safe evacuation horizontally to another 'compartment' from which (if necessary) orderly evacuation can take place in a smoke and heat-free environment.

Number, location and size of escape routes
In order to satisfy the basic principles defined earlier, escape routes should be so planned that any person confronted by an outbreak of fire within a building can turn away from it and make a safe escape. The size and number of escape routes will be determined by the number of people to be evacuated and the need to conform to limitations of distance of travel from any point to a storey exit. As a

rule, therefore, a minimum of two stairs is necessary, but in small premises of limited height and in certain designs of blocks of flats/maisonettes a single stair is acceptable.

The Fire Grading Committee considered 2^1/$_2$ minutes to be the maximum time for evacuation to a place of safety. This time was based on observations of evacuation times in actual fires and the consensus of recommendations in various codes. In the case of upper storeys, the time was taken as being that necessary for evacuation of each storey to a protected stairway and not to the final exit. With modern buildings the units of exit width to achieve this (ie 40 persons per unit width* per minute) are still used. In the case of high-rise buildings, for example, the time needed to evacuate just the floors immediately at risk to a final exit, let alone the evacuation of the entire building, will be much longer.

Minimum widths for exits and stairs are given in the different rules, codes and guides. In general, they are required to be not less than 1100 mm wide (ie 2 units wide) with an incremental increase for more than 220 persons. Reduced exit widths are allowed where small numbers of people are involved. Examples of stair widths for the total evacuation of new office buildings and shops are given in Table 12. However, a single stair width of 1000 mm is considered adequate in buildings comprising flats and/or maisonettes as each dwelling forms a fire compartment and special smoke control measures are provided.

Except where the actual number of persons is known (such as in theatres and restaurants), exit widths are calculated on the basis of the estimated occupant capacity for the particular use of the room or storey. Examples of floor space factors given in BS 5588 are:

Room/Storey	**Floor space factor**[†] (m^2)
Bars	0.3
Committee rooms and lounges	1.0
Shops (high customer density)	2.0
Offices (open plan)	5.0
Other shops and offices	7.0
Storage accommodation	30.0

If not otherwise specified, the treads, risers, going, etc of stairs should be designed in accordance with the appropriate Part of BS 5395. Unenclosed external escape routes are not generally permitted in new buildings but may be the only practicable way of providing alternative means of escape from existing premises.

---

* 500 mm in BS 5588 Parts 2 and 3; 530 mm in connection with Part E of the Building Standards (Scotland) Regulations.
† Floor space per person (excluding stairway enclosures, lifts and sanitary accommodation).

Table 12   Capacity of a stair for total evacuation of a building*

Number of floors served	Maximum number of persons accommodated on one stair of width:				
	1000 mm	1100 mm	1200 mm	1300 mm	1400 mm
1	150	220	240	260	280
2	190	260	285	310	335
3	230	300	330	360	390
4	270	340	375	410	445
5	310	380	420	460	500

*From BS 5588: Part 3

Table 13   Examples of maximum travel distances permitted

	New buildings			Existing buildings
Occupancy	BS 5588	England & Wales[1]	Scotland[2]	Home Office[3]
Factories				
low risk	—	—	45 m (18)	60 m (45)
normal risk	—	45 m (25)	45 m (18)	45 m (25)
high risk	—	25 m (12)	45 m (18)	25 m (12)
Hotels				
bedrooms	—	18 m (9)	—	15 m (8)
bedroom corridors	—	35 m (9)	32 m (15)	35 m (18)
Offices	45 m (18)	45 m (18)	45 m (18)	45 m (18)
Shops				
small[†] ground floor shop	— (27)	—*	—	—
shops which are not small	45 m (18)	45 m (18)	32 m (15)	30 m (18)[‡]
Warehouses				
normal risk	—	45 m (18)	45 m (18)	—
high risk	—	25 m (12)	32 m (15)	—

1   Approved Document B
2   Technical Standards
3   Guides to the Fire Precautions Act 1971
( )   Maximum distance in 'dead ends'
*   Reference is made to BS 5588: Part 2
†   Maximum floor area 280 m^2
‡   Normal risk

Travel distances
These are important both from the point of view of general circulation under fire conditions and in connection with the safety of those people who have to pass through a smoky atmosphere. The maximum distances (see Table 13) over which people should have to travel before reaching a protected stairway or final exit depend on:

(a) whether escape can be in substantially different directions, or in one direction only (ie 'dead end' situations),
(b) the layout of the floor areas within the building, ie whether 'open plan' or subdivided,
(c) the particular occupancy of the building concerned, ie whether assembly, trade, commercial or industrial ('day risk') or residential ('sleeping risk'), and
(d) the physical and/or mental capabilities of the occupants who may need to use the escape routes (eg persons who are elderly, disabled or mentally handicapped).

It should be stressed that travel distances are not related to evacuation times.

Protection against spread of smoke and heat
In the early stages of a fire, the most important effects will usually be those of smoke and other products of combustion. Smoke will be often the first evidence of fire detectable by the occupants in the building and is thus likely to be the first cause of alarm. When first present, smoke tends, in the absence of strong air currents, to collect at ceiling level, filling the space from the top downwards. When it extends down to head height it will produce discomfort to the eyes and difficulty in breathing, which will interfere with the efforts of occupants to find their way towards the exits. People who are prevented from escaping by dense smoke, or who are unduly retarded from escaping by it, may suffer from the heat and toxic effects of the products of combustion within the smoke and the asphyxiant effect caused by the lack of oxygen. Intoxication, incapacity, unconsciousness and possibly death may result. Generally, effective means of escape can therefore only be achieved by the adequate protection of escape routes from the effects of smoke and heat.

Except in two-storey premises with very short travel distances, all escape stairs are required to be enclosed with walls affording not less than a half-hour period of fire resistance and the doors giving access to such 'protected stairways' to be fire-resisting and self-closing. In general, corridors are not required to be similarly enclosed except when forming 'dead ends'. However, there may be situations where the partitions forming the corridor need to be fire-resisting for other purposes, eg if they also form compartment walls.

In certain circumstances therefore, the number of people involved, the type of risk and the distance to a place of safety clear of the premises in the open air, may be such that protection of escape routes is considered unnecessary. In other situations (as with high-rise buildings and buildings, or parts of buildings, served by one stair only) greater emphasis is needed in maintaining the escape routes free from smoke and heat. In these cases the protected stairways are normally required to be entered only through a protected lobby or corridor.

Doors across escape routes
Investigations of many fires in which large numbers of people have died have shown that on many occasions the escape doors were locked or otherwise obstructed. Therefore it is essential that all doors through which the occupants need to pass when making escape from fire must be easily opened in the direction of escape. Where necessary for security reasons, special types of locks are available.

As explained in Chapter 5, any fire door will only fulfil its role if it is fully closed — or is able to close automatically — at the time of a fire. Any attempt by the building users to prevent fire doors from closing may have disastrous effects should a fire occur. Those responsible for the design and maintenance of means of escape should therefore ensure that all fire doors are fitted with the most appropriate type of closer or system for the particular needs of the case.

Signposting and lighting of escape routes
Ideally, all escape routes should be used for normal circulation but this is not always possible or convenient. In many buildings lifts may be the general means of communication between floors, and in other cases emergency exits may not be immediately apparent. It is therefore essential that the occupants are able to easily and quickly identify the location of exits — particularly where the occupants may not be familiar with the layout of the building. Following the principles laid down in BS 5378[149], BS 5499: Part 1[150] specifies that:

safe condition signs (concerning escape routes) should be rectangular with a white symbol or wording on a green background;
prohibition signs should be circular with a black symbol on a white background with a red oblique crossbar and red circumference;
warning signs should be triangular with a black symbol or wording on a yellow background with a black triangular band;
mandatory signs should be circular with a white symbol or wording on a blue background; and
fire equipment signs should be rectangular with a white symbol or wording on a red background.

Figure 23 on the following page shows examples of the different signs.

Recognising the need to cater for a system of signs which can be identified internationally, ISO have published a standard of symbols (ISO 6309[151]), and BS 5499: Part 1 has been revised to take account of the ISO symbols. (Regulations will be made in due course to implement a recent European Council Directive relating to safety signs at work[152] which includes certain fire safety signs.)

Similarly, special provision must be made for the emergency lighting of escape routes in certain situations to ensure that the users can see their way to safety should the main electricity supply fail. It should also be possible for them to see any directional or warning signs, changes in floor level, fire alarm manual call points, etc. Emergency lighting systems should conform to BS 5266: Part 1[153].

a) A warning sign (meaning nearby material is a fire risk)

b) A mandatory sign

c) A prohibition sign (meaning smoking is prohibited)

d) A fire equipment sign

e) Safe condition sign

Figure 23
Examples of fire safety signs

References

1 **Home Office.** *Fire Statistics United Kingdom 1991.* London, HO, 1993.
2 **Ministry of Works.** Fire Grading of Buildings. Part I. General Principles and Structural Precautions. *Post-War Building Studies* No 20. London, HMSO, 1946.
3 **British Standards Institution.** Glossary of terms associated with fire. General terms and phenomena of fire. *British Standard* BS 4422: Part 1: 1987. London, BSI, 1987.
4 **British Standards Institution.** Fire tests on building materials and structures. Test methods and criteria for the fire resistance of elements of building construction. *British Standard* BS 476: Part 8: 1972. London, BSI, 1972.
5 **International Organization for Standardization and the International Electrotechnical Commission.** Glossary of fire terms and definitions. *ISO/IEC Guide* 52: 1990. Geneva, ISO, 1990.
6 **British Standards Institution.** Memo to manufacturers — writing about fire characteristics of materials, products, structures or systems in trade and advertising literature. London, BSI, 1976.
7 **British Standards Institution.** Memo to manufacturers. Don't play with words when writing about fire. London, BSI, 1982.
8 **British Standards Institution.** Memo to manufacturers. Don't be a flaming liability. London, BSI, 1988.
9 **National Council of Building Material Producers.** Can your claims for fire test performance be justified? London, NCBMP.
10 **British Standards Institution.** Glossary of terms relating to burning behaviour of textiles and textile products. *British Standard* BS 6373: 1985. London, BSI, 1985.
11 **British Standards Institution.** Guidelines for the development and presentation of fire tests and for their use in hazard assessment. Draft for development DD64: 1979. London, BSI, 1979.
12 **British Standards Institution.** Guide to development and presentation of fire tests and their use in hazard assessment. *British Standard* BS 6336: 1982. London, BSI, 1982.
13 **Shepherd A F.** Links with the past. London. The Eagle and British Dominions Insurance Company Ltd, 1917.
14 **Jones E L, Porter S and Turner M.** A Gazetteer of English Urban Fire Disasters, 1500–1900. *Historical Geography Research Series.* Number 13. Norwich, Geo Books, 1984.
15 **House of Commons.** Report of the Select Committee on Fire Protection, 1867. HC paper 471.

16 **Metropolitan Board of Works.** The Metropolis Management and Building Acts Amendment Act 1878. Regulations made by the Metropolitan Board of Works at a Meeting held at the Offices of the Board, Spring Gardens, on Friday the 2nd of May 1879, under the provisions of the above-mentioned Act, London, 1879.
17 **Report** of the Royal Commission on Fire Brigades and Fire Prevention. London, HMSO, 1923 (Cmd 1945).
18 **Home Office.** Manual of Safety Requirements in Theatres and other places of Public Entertainment. London, HMSO, 1935.
19 **Fire Protection Association.** *FPA Journal* **35** (October 1956), **54** (October 1961), **56** (July 1962) and **89** (December 1970), and *Fire Prevention* **141** (April 1981) and **181** (July/August 1985). London, FPA.
20 **Housing, England and Wales.** The Housing (Means of Escape from Fire in Houses in Multiple Occupation) Order 1981. *Statutory Instrument* 1981 No 1576. London, HMSO, 1981.
21 **Home Office.** Committee of Inquiry into Crowd Safety and Control at Sports Grounds. Final Report. London, HMSO, 1986 (Cmnd 9710).
22 **Ministry of Works.** Fire Grading of Buildings. Part II Fire fighting equipment, Part III Personal Safety and Part IV Chimneys and flues. *Post-War Building Studies* No 29. London, HMSO, 1952.
23 **Building Industries National Council.** Report on Means of Escape in Case of Fire. London, BINC, 1945.
24 **Building Industries National Council.** Report on Means of Escape from Fire. London, BINC, 1935
25 **British Standards Institution.** Code of functional requirements of buildings. Precautions against fire (Houses and flats of not more than two storeys). *British Standard Code of Practice* CP 3: Chapter IV (1948). London, BSI, 1948.
26 **British Standards Institution.** Code of basic data for the design of buildings. Precautions against fire. Fire precautions in flats and maisonettes over 80 ft in height. *British Standard Code of Practice* CP 3: Chapter IV: Part 1: 1962. London, BSI, 1962.
27 **British Standards Institution.** Code of basic data for the design of buildings. Precautions against fire. Shops and departmental stores. *British Standard Code of Practice* CP 3: Chapter IV: Part 2: 1968. London, BSI, 1968.
28 **British Standards Institution.** Code of basic data for the design of buildings. Precautions against fire. Office buildings. *British Standard Code of Practice* CP 3: Chapter IV: Part 3: 1968. London, BSI, 1968.
29 **British Standards Institution.** Code of basic data for the design of buildings. Precautions against fire. Flats and maisonettes (in blocks over two storeys). *British Standard Code of Practice* CP3: Chapter IV: Part 1: 1971. London, BSI, 1971.
30 **British Standards Institution.** Fire precautions in the design, construction and use of buildings. Code of practice for residential buildings. *British Standard* BS 5588: Part 1: 1990. London, BSI, 1990.
31 **British Standards Institution.** Fire precautions in the design and construction of buildings. Code of practice for shops. *British Standard* BS 5588: Part 2: 1985. London, BSI, 1985.

32 **British Standards Institution.** Fire precautions in the design and construction of buildings. Code of practice for office buildings. *British Standard* BS 5588: Part 3: 1983. London, BSI, 1983.
33 **Holland H.** Resolutions of the Associated Architects; with the report of a committee by them appointed to consider the causes of the frequent fires, and the best means of preventing the like in future. London, 1793.
34 **Davey N and Ashton L A.** Investigations on Building Fires. Part V. Fire Tests on Structural Elements. *National Building Studies Research Paper* No 12. London, HMSO, 1953.
35 **The British Fire Prevention Committee.** The Standards of Fire Resistance of the British Fire Prevention Committee (as adopted to serve as universal standards at the International Fire Prevention Congress, London, 1903). *Publication* 82. London, BFPC, 1904.
36 **National Fire Brigades Association.** *Red book* 253. London, NFBA, 1929.
37 **British Standards Institution.** British Standard Definitions for Fire-resistance, Incombustibility and Non-inflammability of Building Materials and Structures (including methods of test). *British Standard* BS 476: 1932. London, BSI, 1932.
38 **British Standards Institution.** Fire tests on building materials and structures. *British Standard* BS 476: 1953. London, BSI, 1953.
39 **British Standards Institution.** Fire tests on building materials and structures. Flammability test for thin flexible materials. *British Standard* BS 476: Part 2: 1955. London, BSI, 1955.
40 **British Standards Institution.** Fire tests on building materials and structures. Non-combustibility test for materials. *British Standard* BS 476: Part 4: 1970. London, BSI, 1970.
41 **British Standards Institution.** Fire tests on building materials and structures. Surface spread of flame tests for materials. *British Standard* BS 476: Part 7: 1971. London, BSI, 1971.
42 **British Standards Institution.** Fire tests on building materials and structures. Fire propagation test for materials. *British Standard* BS 476: Part 6: 1968. London, BSI, 1968.
43 **Rogowski, Barbara F W.** The fire propagation test: its development and application. *Fire Research Technical Paper* No 25. London, HMSO, 1970.
44 **Webster C T.** Development of a roof test. *Fire Research Station Fire Research Note* 289. FRS, Borehamwood, 1957.
45 **British Standards Institution.** Fire tests on building materials and structures. External fire exposure roof tests. *British Standard* BS 476: Part 3: 1958. London, BSI, 1958.
46 **Home Office.** Review of fire policy: an examination of the deployment of resources to combat fire. London, HO, 1980.
47 **Pitt P H.** Guide to building control by local acts 1987. London, Architectural Press, 1987.
48 **Department of the Environment and Welsh Office.** Building Regulations 1972/3. Guidance Note. Structural fire precautions. London, HMSO, 1975.
49 **Department of the Environment and The Welsh Office.** The Building Regulations 1985. Mandatory rules for means of escape in case of fire. London, HMSO, 1985.

50 **Malhotra H L.** *Fire safety in buildings.* Building Research Establishment Report. Garston, BRE, 1986.
51 **Building and Buildings.** The Building Regulations 1991. *Statutory Instrument* 1991 No 2768. London, HMSO, 1991.
52 **Department of the Environment and The Welsh Office.** *The Building Regulations 1991. Approved Document B. Fire safety* (1992 edition). London, HMSO, 1991.
53 **Building and Buildings.** The Building Standards (Scotland) Regulations 1990. *Statutory Instrument* 1990 No 2179 (S.187). Edinburgh, HMSO, 1990.
54 **The Scottish Office.** *The Building Standards (Scotland) Regulations 1990. Technical Standards for compliance with the Building Standards (Scotland) Regulations 1990.* Edinburgh, HMSO, 1990.
55 **Building Regulations.** The Building Regulations (Northern Ireland) 1990. *Statutory Rules of Northern Ireland* 1990 No 59. Belfast, HMSO, 1990.
56 **Building and Buildings.** The Building (Inner London) Regulations 1985. *Statutory Instrument* 1985 No 1936. London, HMSO, 1985.
57 **European Economic Community.** Council Directive of 21 December 1988 on the approximation of laws, regulations and administrative provisions of the member States relating to construction products (89/106/EEC).
58 **Building and Buildings.** The Construction Products Regulations 1991. *Statutory Instrument* 1991 No 1620. London, HMSO, 1991.
59 **Department of Trade and Industry.** *Fire and Building Regulation.* A Review by Bickerdike Allen Partners. London, HMSO, 1990.
60 **Department of the Environment, Home Office, Welsh Office.** Building Regulation and Fire Safety. Procedural guidance. London, DOE, 1992.
61 **Fire Precautions.** Sports Grounds and Sporting Events. The Fire Safety and Safety of Places of Sport Act 1987 (Commencement No 1) Order 1987. *Statutory Instrument* 1987 No 1762 (C.54). London, HMSO, 1987.
62 **Fire Precautions.** The Fire Safety and Safety of Places of Sport Act 1987 (Commencement No 5) Order 1989. *Statutory Instrument* 1989 No 75 (C.3). London, HMSO, 1989.
63 **Fire Precautions.** The Fire Precautions (Hotels and Boarding Houses) Order 1972. *Statutory Instrument* 1972 No 238. London, HMSO, 1972.
64 **Fire Precautions.** The Fire Precautions (Hotels and Boarding Houses) (Scotland) Order 1972. *Statutory Instrument* 1972 No 382 (S.26). Edinburgh, HMSO, 1972.
65 **Fire Precautions.** The Fire Precautions (Factories, Offices, Shops and Railway Premises) Order 1976. *Statutory Instrument* 1976 No 2009. London, HMSO, 1976.
66 **Fire Precautions.** The Fire Precautions (Factories, Offices, Shops and Railway Premises) Order 1989. *Statutory Instrument* 1989 No 76. London, HMSO, 1989.
67 **Home Office/The Scottish Office.** Fire Precautions Act 1971. Guide to fire precautions in premises used as hotels and boarding houses which require a fire certificate. London, HMSO, 1991.
68 **Home Office/Scottish Home and Health Department.** Fire Precautions Act 1971 Guide to fire precautions in existing places of work that require a fire certificate. Factories, Offices, Shops and Railway Premises. London, HMSO, 1989.

69 **Home Office/Scottish Home and Health Department.** Guides to the Fire Precautions Act 1971. 1 Hotels and Boarding Houses. London, HMSO, 1972.
70 **Home Office/Scottish Home and Health Department.** Guides to the Fire Precautions Act 1971. 2 Factories. London, HMSO, 1977.
71 **Home Office/Scottish Home and Health Department.** Guides to the Fire Precautions Act 1971. 3 Offices, Shops and Railway Premises. London, HMSO, 1977.
72 **Home Office/The Scottish Office.** Fire Precautions Act 1971. Fire safety management in hotels and boarding houses. London, HMSO, 1991.
73 **Home Office/Scottish Home and Health Department.** Fire Precautions Act 1971. Fire safety at work. London, HMSO, 1989.
74 **National Health Service, England and Wales. National Health Service, Scotland.** The National Health Service and Community Care Act 1990 (Commencement No 1) Order 1990. *Statutory Instrument* 1990 No 1329 (C.37). London, HMSO, 1990.
75 **Northern Ireland.** The Fire Services (Northern Ireland) Order 1984. *Statutory Instrument* 1984 No 1821 (NI.11). Belfast, HMSO, 1984.
76 **Fire Services.** The Fire Services (Hotels and Boarding Houses) Order (Northern Ireland) 1985. *Statutory Rules of Northern Ireland* 1985 No 138. Belfast, HMSO, 1985.
77 **Fire Services.** The Fire Services (Factory, Office and Shop Premises) Order (Northern Ireland) 1986. *Statutory Rules of Northern Ireland* 1986 No 355. Belfast, HMSO, 1986.
78 **Fire Services.** The Fire Services (Leisure Premises) Order (Northern Ireland) 1985. *Statutory Rules of Northern Ireland* 1985 No 137. Belfast, HMSO, 1985.
79 **Fire Services.** The Fire Services (Betting, Gaming and Amusement Premises) Order (Northern Ireland) 1987. *Statutory Rules of Northern Ireland* 1987 No 334. Belfast, HMSO, 1987.
80 **Home Office/Scottish Home and Health Department.** Code of practice for fire precautions in factories, offices, shops and railway premises not required to have a fire certificate. London, HMSO, 1989.
81 **Fire Precautions.** The Fire Precautions (Non-Certificated Factory, Office, Shop and Railway Premises) Regulations 1976. *Statutory Instrument* 1976 No 2010. London, HMSO, 1976.
82 **Fire Services.** The Fire Services (Non-Certificated Factory, Office and Shop Premises) Regulations (Northern Ireland) 1986. *Statutory Rules of Northern Ireland* 1986 No 352. Belfast, HMSO, 1986.
83 **Fire Precautions.** The Fire Precautions (Sub-surface Railway Stations) Regulations 1989. *Statutory Instrument* 1989 No 1401. London, HMSO, 1989.
84 **European Economic Community.** Council Directive of 12 June 1989 on the introduction of measures to encourage improvements in the safety and health of workers at work. (89/391/EEC).
85 **European Economic Community.** Council Directive of 30 November 1989 concerning the minimum safety and health requirements for the workplace (first individual directive within the meaning of Article 16 (1) of Directive 89/391/EEC) (89/654/EEC).
86 **Health and Safety.** Fire Certificates (Special Premises) Regulations 1976. *Statutory Instrument* 1976 No 2003. London, HMSO, 1976.

87 **Health and Safety.** Fire Certificates (Special Premises) Regulations (Northern Ireland) 1991. *Statutory Rules of Northern Ireland* 1991 No 446. Belfast, HMSO, 1991.

88 **Health and Safety Executive.** 20 Questions. The Fire Certificates (Special Premises) Regulations 1976. London, HMSO, 1976.

89 **Health and Safety Executive.** Guidance on general fire precautions at premises subject to the Fire Certificates (Special Premises) Regulations 1976. London, HSE, 1982.

90 **Department of the Environment, Home Office and Welsh Office.** Memorandum on overcrowding and Houses in Multiple Occupation. The Housing Act 1985: Parts X, XI and XII. *Joint Circular* 12/86 (DOE), 39/86 (HO) and 23/86 (WO). London, HMSO, 1986.

91 **Department of the Environment and Welsh Office.** Houses in Multiple Occupation. Local Government and Housing Act 1989: Parts VII and VIII, Schedules 9 and 11. *Joint Circular* 5/90 (DOE) and 17/90 (WO). London, HMSO, 1990.

92 **Department of the Environment.** Houses in Multiple Occupation. Guidance to local housing authorities on standards of fitness under section 352 of the Housing Act 1985. *Circular* 12/92. London, HMSO, 1992.

93 **Home Office/Department of the Environment/Welsh Office.** Guide to means of escape and related fire safety measures in certain existing houses in multiple occupation. London, HMSO, 1988.

94 **Scottish Home and Health Department.** Guide to means of escape and related fire safety measures in existing houses in multiple occupation in Scotland. Edinburgh, HMSO, 1988.

95 **Loss Prevention Council.** Code of practice for the construction of buildings. London, LPC, 1992.

96 **Hilado C J.** Flammability Handbook for Plastics. Stamford Conn. USA, Technomic, 1969.

97 **British Standards Institution.** Fire tests on building materials and structures. Method for classification of the surface spread of flame of products. *British Standard* BS 476: Part 7: 1987. London, BSI, 1987.

98 **British Standards Institution.** Fire tests on building materials and structures. Method of test for fire propagation for products. *British Standard* BS 476: Part 6: 1989. London, BSI, 1989.

99 **British Standards Institution.** Fire tests on building materials and structures. Method for determination of the fire resistance of elements of construction (general principles). *British Standard* BS 476: Part 20: 1987. London, BSI, 1987.

100 **British Standards Institution.** Fire tests on building materials and structures. Methods for determination of the fire resistance of loadbearing elements of construction. *British Standard* BS 476: Part 21: 1987. London, BSI, 1987.

101 **British Standards Institution.** Fire tests on building materials and structures. Methods for determination of the fire resistance of non-loadbearing elements of construction. *British Standard* BS 476: Part 22: 1987. London, BSI, 1987.

102 **British Standards Institution.** Fire tests on building materials and structures. Methods for determination of the contribution of components to the fire resistance of a structure. *British Standard* BS 476: Part 23: 1987. London, BSI, 1987.

103 **British Standards Institution.** Fire tests on building materials and structures. Method for determination of the fire resistance of ventilation ducts. *British Standard* BS 476: Part 24: 1987. London, BSI, 1987.

104 **British Standards Institution.** Glossary of terms associated with fire. Structural fire protection. *British Standard* BS 4422: Part 2: 1990. London, BSI, 1990.

105 **Chitty T B, Nicholson D and Malhotra H L.** Tests on roof constructions subjected to external fire. *Fire Note* No 4. London, HMSO, 1970.

106 **Fisher R W and Smart P M T.** *Results of fire resistance tests on elements of building construction.* Volumes 1 and 2. Building Research Establishment Reports. London, HMSO, 1975 and 1976.

107 **Fisher R W and Rogowski, Barbara F W.** *Results of fire propagation tests on building products.* Building Research Establishment Report. London, HMSO, 1976.

108 **Fisher R W and Rogowski, Barbara F W.** *Results of surface spread of flame tests on building products.* Building Research Establishment Report. London, HMSO, 1976.

109 **Fisher R W.** *Fire tests results on building products. Surface spread of flame.* London, Fire Protection Association, 1979 (updated 1981).

110 **Fisher R W.** *Fire tests results on building products. Fire propagation.* London, Fire Protection Association, 1980 (updated 1986).

111 **Fisher R W.** *Fire tests results on building products. Fire resistance.* London, Fire Protection Association, 1983.

112 **British Standards Institution.** Fire tests on building materials and structures. Method of test for ignitability. *British Standard* BS 476: Part 5: 1979. London, BSI, 1979.

113 **British Standards Institution.** Fire tests on building materials and structures. Method for assessing the heat emission from building materials. *British Standard* BS 476: Part 11: 1982. London, BSI, 1982.

114 **British Standards Institution.** Methods of test for flammability of textile fabrics when subjected to a small igniting flame applied to the face or bottom edge of vertically oriented specimens. *British Standard* BS 5438: 1989. London, BSI, 1989.

115 **Fire Protection Association.** Flame retardant treatments for wood and its derivatives. Fire and the Architect. *Fire Prevention Design Guide* No 13. London, FPA, 1977.

116 **Timber Research and Development Association.** Flame retardant treatments for timber. *TRADA Wood Information.* High Wycombe, TRADA, 1982.

117 **Morris W A, Read R E H and Cooke G M E.** *Guidelines for the construction of fire-resisting structural elements.* Building Research Establishment Report. Garston, BRE, 1988.

118 **The Institution of Structural Engineers.** Report of a Joint Committee of the Institution of Structural Engineers and the Concrete Society. Fire resistance of concrete structures. London, ISE, 1975.

119 **Association of Specialist Fire Protection Contractors and Manufacturers.** *Fire protection for structural steel in buildings.* Aldershot, ASFPCM, 1992, 2nd edition (revised).

120 **Cooke G M E.** New methods of fire protection for external steelwork. *The Architects Journal,* 1974, **160** (35) 494–516.

121 **Cooke G M E.** Water filled structures. British Steel Corporation (Tubes Division), August 1973.
122 **British Standards Institution.** Structural use of timber. Fire resistance of timber structures. Recommendations for calculating fire resistance of timber members. *British Standard* BS 476: Part 4: Section 4.1: 1978. London, BSI, 1978.
123 **Malhotra H L.** *Design of fire-resisting structures.* Glasgow, Surrey University Press, 1982.
124 **British Standards Institution.** Windows and rooflights. Fire hazards associated with glazing in buildings. *British Standard Code of Practice* CP 153: Part 4: 1972. London, BSI, 1972.
125 **British Standards Institution.** Use of elements of structural fire protection with particular reference to the recommendations given in BS 5588 'Fire precautions in the design and construction of buildings'. Guide to the fire performance of glass. *Published Document* PD 6512: Part 3: 1987. London, BSI, 1987.
126 **British Standards Institution.** Structural use of timber. Fire resistance of timber structures. Recommendations for calculating fire resistance of timber stud walls and joisted floor constructions. *British Standard* BS 476: Part 4: Section 4.2: 1990. London, BSI, 1990.
127 **Newman G M.** *Fire and steel construction: the behaviour of steel portal frames in boundary conditions.* Ascot, The Steel Construction Institute, 1990, 2nd edition.
128 **Society for the Protection of Ancient Buildings and Council for Small Industries in Rural Areas.** The care and repair of thatched roofs. *Technical pamphlet* No 10. (Available from the Rural Development Commission, 141 Castle Street, Salisbury, Wilts, SP1 3TP).
129 **British Standards Institution.** Use of clements of structural fire protection with particular reference to the recommendations given in BS 5588 'Fire precautions in the design and construction of buildings'. Guide to fire doors. *Published Document* PD 6512: Part 1: 1985. London, BSI, 1985.
130 **British Standards Institution.** Fire tests on building materials and structures. Methods for measuring smoke penetration through doorsets and shutter assemblies. Method of measurement under ambient temperature conditions. *British Standard* BS 476: Part 31: Section 31.1: 1983. London, BSI, 1983.
131 **British Standards Institution.** Fire-check flush doors and wood and metal frames (half-hour and one-hour types). *British Standard* BS 459: Part 3: 1951. London, BSI, 1951.
132 **British Standards Institution.** Code of practice for fire door assemblies with non-metallic leaves. *British Standard* BS 8214: 1990. London, BSI, 1990.
133 **British Standards Institution.** Fire detection and alarm systems for buildings. Specification for automatic release mechanisms for certain fire protection equipment. *British Standard* BS 5839: Part 3: 1988. London, BSI, 1988.
134 **Langdon-Thomas G J and Hinkley P L.** Fire venting in single-storey buildings. *Fire Note* No 5. London, HMSO, 1965.
135 **Morgan H P and Gardner J P.** *Design principles for smoke ventilation in enclosed shopping centres.* Building Research Establishment Report. Garston, BRE, 1990.

136 **Hinkley P L and Theobald C R.** PVC rooflights for venting fires in single-storey buildings. *Fire Research Technical Paper* No 14. London, HMSO, 1966.

137 **British Standards Institution.** Fire precautions in the design and construction of buildings. Code of practice for means of escape for disabled people. *British Standard* BS 5588: Part 8: 1988. London, BSI, 1988.

138 **British Standards Institution.** Glossary of terms associated with fire. Evacuation and means of escape. *British Standard* BS 4422: Part 6: 1988. London, BSI, 1988.

139 **Department of the Environment for Northern Ireland.** The Building Regulations (Northern Ireland) 1990. *Technical Booklet* E. Fire. Belfast, HMSO, [1993].

140 **British Standards Institution.** Fire precautions in the design, construction and use of buildings. Code of practice for places of assembly. *British Standard* BS 5588: Part 6: 1991. London, BSI, 1991.

141 **British Standards Institution.** Fire precautions in the design, construction and use of buildings. Code of practice for shopping complexes. *British Standard* BS 5588: Part 10: 1991. London, BSI, 1991.

142 **Department of Education and Science.** Fire and the design of educational buildings. *Building Bulletin* 7. London, HMSO, 1988.

143 **Department of Health and Social Security and the Welsh Office.** *Firecode.* Fire precautions in new hospitals. HTM 81. London, HMSO, 1987.

144 **Department of Health and the Welsh Office.** *Firecode.* Nucleus fire precautions recommendations. London, HMSO, 1989.

145 **Scottish Home and Health Department.** Fire safety. New health buildings in Scotland. Edinburgh, Scottish Office, 1987.

146 **Home Office/Scottish Home and Health Department.** Guide to fire precautions in existing places of entertainment and like premises. London, HMSO, 1990.

147 **Home Office/Scottish Home and Health Department.** Draft guide to fire precautions in hospitals. London, Home Office, 1982.

148 **Home Office/Scottish Home and Health Department.** Draft guide to fire precautions in existing residential care premises. London, Home Office, 1983.

149 **British Standards Institution.** Safety signs and colours. Specification for colour and design. *British Standard* BS 5378: Part 1: 1980. London, BSI, 1980.

150 **British Standards Institution.** Fire safety signs, notices and graphic symbols. Specification for fire safety signs. *British Standard* BS 5499: Part 1: 1990. London, BSI, 1990.

151 **International Organization for Standardization.** Fire protection — Safety signs. ISO 6309: 1987. Geneva, ISO, 1987.

152 **European Community.** Council Directive of 24 June 1992 on the minimum requirements for the provision of safety and/or health signs at work. (92/58/EEC).

153 **British Standards Institution.** Emergency lighting. Code of practice for the emergency lighting of premises other than cinemas and certain other specified premises used for entertainment. *British Standard* BS 5266: Part 1: 1988. London, BSI, 1988.

Appendix 1: Extracts from historical documents

A. London Building Act 1894 (The Second Schedule)

The following materials shall for the purposes of this Act be deemed to be fire-resisting materials:—

(1) Brickwork constructed of good bricks well burnt hard and sound properly bonded and solidly put together —
   (a) With good mortar compounded of good lime and sharp clean sand hard clean broken brick broken flint grit or slag; or
   (b) With good cement; or
   (c) With cement mixed with sharp clean sand hard clean broken brick broken flint grit or slag:

(2) Granite and other stone suitable for building purposes by reason of its solidity and durability:

(3) Iron steel and copper:

(4) Oak and teak and other hard timber when used for beams or posts or in combination with iron the timber and the iron (if any) being protected by plastering in cement or other incombustible or non-conducting external coating;
In the case of doors —
oak or teak or other hard timber not less than two inches thick:
In the case of staircases —
oak or teak or other hard timber with treads strings and risers not less than two inches thick:

(5) Slate tiles brick and terra cotta when used for coverings or corbels:

(6) Flagstones when used for floors over arches but not exposed on the underside and not supported at the ends only:

(7) Concrete composed of broken brick stone chippings or ballast and lime cement or calcined gypsum when used for filling in between joists of floors:

(8) Any material from time to time approved by the Council as fire-resisting.

B. Standards of fire resistance of the British Fire Prevention Committee

Standard table for fire-resisting floors and ceilings

Classification	Sub-class	Duration of test at least	Minimum temperature	Load per superficial foot distributed *(per square metre)*	Minimum superficial area under test	Minimum time for application of water under pressure
Temporary protection	Class A	45 min	1500°F (815.5°C)	Optional	100 sq ft (9.290 sq m)	2 min
	Class B	60 min	1500°F (815.5°C)	Optional	200 sq ft (18.580 sq m)	2 min
Partial protection	Class A	90 min	1800°F (982.2°C)	112 lbs (546.852 kg)	100 sq ft (9.290 sq m)	2 min
	Class B	120 min	1800°F (982.2°C)	168 lbs (820.278 kg)	200 sq ft (18.580 sq m)	2 min
Full protection	Class A	150 min	1800°F (982.2°C)	224 lbs (1093.706 kg)	100 sq ft (9.290 sq m)	2 min
	Class B	240 min	1800°F (982.2°C)	280 lbs (1367.130 kg)	200 sq ft (18.580 sq m)	5 min

Standard table for fire-resisting partitions

Classification	Sub-class	Duration of test at least	Minimum temperature	Thickness of material	Minimum superficial area under test	Minimum time for application of water under pressure
Temporary protection	Class A	45 min	1500°F (815.5°C)	2 in & under (.051 m)	80 sq ft (7.432 sq m)	2 min
	Class B	60 min	1500°F (815.5°C)	Optional	80 sq ft (7.432 sq m)	2 min
Partial protection	Class A	90 min	1800°F (982.2°C)	2½ in & under (.063 m)	80 sq ft (7.432 sq m)	2 min
	Class B	120 min	1800°F (982.2°C)	Optional	80 sq ft (7.432 sq m)	2 min
Full protection	Class A	150 min	1800°F (982.2°C)	2½ in & under (.063 m)	80 sq ft (7.432 sq m)	2 min
	Class B	240 min	1800°F (982.2°C)	Optional	80 sq ft (7.432 sq m)	5 min

Standard table for fire-resisting single doors, with or without frames

Classification	Sub-class	Duration of test at least	Minimum temperature	Thickness of material	Minimum superficial area under test	Minimum time for application of water under pressure
Temporary protection	Class A	45 min	1500°F (815.5°C)	2 in & under (.051 m)	20 sq ft (1.858 sq m)	2 min
Temporary protection	Class B	60 min	1500°F (815.5°C)	Optional	20 sq ft (1.858 sq m)	2 min
Partial protection	Class A	90 min	1800°F (982.2°C)	2½ in & under (.063 m)	20 sq ft (1.858 sq m)	2 min
Partial protection	Class B	120 min	1800°F (982.2°C)	Optional	20 sq ft (1.858 sq m)	2 min
Full protection	Class A	150 min	1800°F (982.2°C)	½ in & under (.018 m)	25 sq ft (2.322 sq m)	2 min
Full protection	Class B	240 min	1800°F (982.2°C)	Optional	25 sq ft (2.322 sq m)	5 min

NOTE:
Temporary Protection implies resistance against fire for at least ¼ hour.
Partial Protection implies resistance against a fierce fire for at least 1½ hours.
Full Protection implies resistance against a fierce fire for at least 2½ hours.

Appendix 2: Extracts from current legislation

A. London Building Acts (Amendment) Act 1939*

Section 20. 'Precautions against fire in certain buildings and cubical extent of buildings'

Applies where:—
(a) a building is to be erected with a storey or part of a storey at a greater height than —
   (i) 30 metres; or
   (ii) 25 metres if the area of the building exceeds 930 square metres;

(b) a building of the warehouse class, or a building or part of a building used for the purposes of trade or manufacture, exceeds 7100 cubic metres in extent unless it is divided by division walls in such manner that no division of the building is of a cubical extent exceeding 7100 cubic metres.

Section 34. 'Protection against fire in certain new[†] buildings'

Applies to:—
Every new public building every new building which is constructed to be used or is used in whole or in part as a church chapel or other place of worship hall meeting room school classroom concert room dancing room or other place of assembly and every other new building:—
(a) which if of one storey exceeds six squares (600 sq ft) in area; or

---

*As amended by the Building (Inner London) Regulations 1985 and 1987.
†Buildings erected after 1.1.1940, and certain 'old' buildings which have been partially demolished and rebuilt, or extended or altered in certain respects.
'building of the warehouse class' means a warehouse manufactory brewery or distillery or any other building exceeding in cubical extent 150,000 cub ft which is neither a public building nor a domestic building.
'domestic building' includes a dwelling-house and any other building not being either a public building or a building of the warehouse class.
'public building' means:—
(a) a building used wholly or partly as a church chapel or other place of public worship (not being a dwelling-house so used) or as a public assistance institution or public library or as a place for public entertainments public balls public dances public lectures or public exhibitions or otherwise as a place of public assembly, or
(b) a building of a cubical extent exceeding 250,000 cub ft which is used wholly or partly as an hotel or hospital or as a school college or other place of instruction.

(b) which if of more than one storey has in the aggregate a total floor area exceeding ten squares (1000 sq ft) (exclusive of any basement storey used solely for storage purposes); or
(c) which has a storey at a greater height than twenty feet; or
(d) in which more than ten persons are employed above the ground storey.

But not if requirements for means of escape are imposed under building regulations; or to a dwelling house which does not contain a storey at a greater height than 20 ft, if the house is essentially used for human habitation and is not constructed to be let in self-contained flats or tenements or in habitable rooms in different occupation, or as an hotel or boarding house; or to buildings covered by section 4(1)(a) of the Building Act 1984.

Section 35. 'Protection against fire in certain old* buildings'

Applies where an old building:—
(a) except a dwelling-house occupied as such by not more than one family
    (i) contains any storey which is at a greater height than forty-two feet; or
    (ii) is a building in which sleeping accommodation is provided for more than twenty persons or which is occupied by more than twenty persons or in which more than twenty persons are employed; or
(b) is a building in which more than ten persons are normally employed at any one time above the first storey or on or above any storey which is at a greater height than twenty feet; or
(c) exceeds two storeys in height and contains any storey which is at a greater height than twenty feet and:—
    (i) is let in flats or tenements; or
    (ii) is used as an inn hotel boarding house hospital nursing home boarding school children's home or other institution; or
    (iii) is used as a restaurant shop store or warehouse and has on any storey above ground storey any sleeping accommodation; or
(d) contains a place of assembly having a superficial area of not less than five hundred square feet.

But not to any building covered by section 4(1)(a) of the Building Act 1984.

B. Fire Precautions Act 1971[†]

Section 1. 'Uses of premises for which fire certificate is compulsory'

(1) A certificate issued under this Act by the fire authority (in this Act referred to as a "fire certificate") shall, subject to any exemption conferred by or under this Act, be required in respect of any premises which are put to a use for the time being designated under this section (in this Act referred to as a "designated use").

---

* Not being a 'new' building.
[†] As amended by the Health and Safety at Work etc Act 1974 and the Fire Safety and Safety of Places of Sport Act 1987.

(2) For the purposes of this section the Secretary of State may by order designate particular uses of premises, but shall not so designate any particular use unless it falls within at least one of the following classes of use, that is to say:—
    (a) use as, or for any purpose involving the provision of, sleeping accommodation;
    (b) use as, or as part of, an institution providing treatment or care;
    (c) use for purposes of entertainment, recreation or instruction or for purposes of any club, society or association;
    (d) use for purposes of teaching, training or research;
    (e) use for any purpose involving access to the premises by members of the public, whether on payment or otherwise;
    (f) use as a place of work.

(3) An order under this section may provide that a fire certificate shall not by virtue of this section be required for premises of any description specified in the order, notwithstanding that they are or form part of premises which are put to a designated use.

Section 2. 'Premises exempt from s.1'

No fire certificate shall by virtue of section 1 of this Act be required in respect of premises consisting of or comprised in a house which is occupied as a single private dwelling.

Section 3. 'Power of fire authority to make fire certificate compulsory for use of certain premises as a dwelling'

Does not apply to single family dwelling houses or to houses in multiple occupation.

Section 4. 'Right of appeal against, and coming into force of, notices under s.3'

Section 5. 'Application for, and issue of, fire certificate'

Section 5A. 'Powers for fire authority to grant exemption in particular cases'

Section 5B. 'Withdrawal of exemptions under s.5A'

Section 6. 'Contents of fire certificate'

(1) Every fire certificate issued with respect to any premises shall specify:—
    (a) the particular use or uses of the premises which the certificate covers; and
    (b) the means of escape in case of fire with which the premises are provided; and
    (c) the means (other than means for fighting fire) with which the relevant building is provided for securing that the means of escape with which

the premises are provided can be safely and effectively used at all material times; and
- (d) the type, number and location of the means for fighting fire (whether in the premises or affecting the means of escape) with which the relevant building is provided [for use in case of fire by persons in the building]*; and
- (e) the type, number and location of the means with which the relevant building is provided for giving to persons in the premises warning in case of fire; and
- (f)† in the case of a factory premises, particulars as to any explosive or highly flammable materials which may be stored or used in the premises;

and may, where appropriate, do so by means of or by reference to a plan.

(2) A fire certificate issued with respect to any premises may impose such requirements as the fire authority consider appropriate in the circumstances:—
- (a) for securing that the means of escape in case of fire with which the premises are provided are properly maintained and kept free from obstruction;
- (b) for securing that the means with which the relevant building is provided as mentioned in subsection (1)(c) to (e) above are properly maintained;
- (c) for securing that persons employed to work in the premises receive appropriate instruction or training in what to do in case of fire, and that records are kept of instruction or training given for that purpose;
- (d) for limiting the number of persons who may be in the premises at any one time; and
- (e) as to other precautions to be observed in the relevant building in relation to the risk, in case of fire, to persons in the premises.

Section 7. 'Offences in relation to foregoing provisions'

Section 8. 'Change of conditions affecting adequacy of certain matters specified in fire certificate, etc'

Section 8B. 'Charges for issue or amendment of fire certificates'

Section 9A. 'Duty as to means of escape and for fighting fire'
(1) All premises to which this section applies shall be provided with —
- (a) such means of escape in case of fire, and
- (b) such means for fighting fire,

as may reasonably be required in the circumstances of the case.

---

* The wording in square brackets will be deleted on 1 August 1993 (SI 1993 No 1411) so as to enable sprinklers or other automatic extinguishing systems to be required as a condition of issuing a certificate.
† Added by the Fire Precautions Act 1971 (Modifications) Regulations 1976. SI 1976 No 2007.

(2) The premises to which this section applies are premises which are exempt from the requirement for a fire certificate by virtue of —
 (a) a provision made in an order under section 1 of this Act by virtue of subsection (3) of that section, or
 (b) the grant of exemption by a fire authority under section 5A of this Act.

Section 9B. 'Codes of practice as to means of escape and for fighting fire'

Section 10. 'Special procedure in case of serious risk: prohibition notices'

Section 10A. 'Rights of appeal against prohibition notices'

Section 10B. 'Provision as to offences'

Section 12. 'Power of Secretary of State to make regulations about fire precautions'

(1) In the case of any particular use of premises which he has power to designate under section 1 of this Act the Secretary of State may by regulations make provision as to the precautions which, as regards premises put to that use, or any specified class of such premises, are to be taken or observed in relation to the risk to persons in case of fire, but so that nothing in any regulations made under this section shall apply to premises of the description given in section 2 of this Act and nothing in this section shall confer on the Secretary of State power to make provision with respect to the taking or observance of special precautions in connection with the carrying on of any manufacturing process.

(2) The Secretary of State may by regulations make provision as to the precautions which are to be taken or observed in relation to the risk to persons in case of fire as regards premises which, while section 3 of this Act applies to them and a notice under that section is in force in relation to them, are used as a dwelling, or any specified class of such premises.

(3) Without prejudice to the generality of the powers conferred on the Secretary of State by subsections (1) and (2) above, regulations made by him under this section may in particular, as regards any premises to which they apply, impose requirements —
 (a) as to the provision, maintenance and keeping free from obstruction of means of escape in case of fire;
 (b) as to the provision and maintenance of means for securing that any means of escape can be safely and effectively used at all material times;
 (c) as to the provision and maintenance of means for fighting fire and means for giving warning in case of fire;
 (d) as to the internal construction of the premises and the materials used in that construction;
 (e) for prohibiting altogether the presence or use in the premises of furniture or equipment of any specified description, or prohibiting its

presence or use unless specified standards or conditions are complied with;
- (f) for securing that persons employed to work in the premises receive appropriate instruction or training in what to do in case of fire;
- (g) for securing that, in specified circumstances, specified numbers of attendants are stationed in specified parts of the premises; and
- (h) as to the keeping of records of instruction or training given, or other things done, in pursuance of the regulations.

(4) Regulations under this section —
- (a) may impose requirements on persons other than occupiers of premises to which they apply; and
- (b) may, as regards any of their provisions, make provision as to the person or persons who are to be responsible for any contravention thereof; and
- (c) may provide that if any specified provision of the regulations is contravened, the person or each of the persons who under the regulations is or are responsible for the contravention shall be guilty of an offence under this section.

Section 13. 'Exercise of certain powers of fire authority in England or Wales where building regulations as to means of escape apply'

Section 14. 'Exercise of certain powers of fire authority in Scotland where building standards regulations as to means of escape apply'

NB. Sections 13 and 14 cover what has been generally termed the 'statutory bar'.

Section 16. 'Duty of local authority to consult fire authority in certain cases before passing plans'.

Section 17. 'Duty of fire authorities to consult other authorities before requiring alterations to buildings'.

C. Fire Certificates (Special Premises) Regulations 1976

Schedule 1, Part I. 'Premises for which a fire certificate is required'

1. Any premises at which are carried on any manufacturing processes in which the total quantity of any highly flammable liquid under pressure greater than atmospheric pressure and above its boiling point at atmospheric pressure may exceed 50 tonnes.

2. Any premises at which is carried on the manufacture of expanded cellular plastics and at which the quantities manufactured are normally of, or in excess of, 50 tonnes per week.

3. Any premises at which there is stored, or there are facilities provided for the storage of, liquified petroleum gas in quantities of, or in excess of, 100 tonnes

except where the liquified petroleum gas is kept for use at the premises either as fuel, or for the production of an atmosphere for the heat-treatment of metals.

4. Any premises at which there is stored, or there are facilities provided for the storage of, liquified natural gas in quantities of, or in excess of, 100 tonnes except where the liquified natural gas is kept solely for use at the premises as a fuel.

5. Any premises at which there is stored, or there are facilities provided for the storage of, any liquified flammable gas consisting predominantly of methyl acetylene in quantities of, or in excess of, 100 tonnes except where the liquified flammable gas is kept solely for use at the premises as a fuel.

6. Any premises at which oxygen is manufactured and at which there are stored, or there are facilities provided for the storage of, quantities of liquid oxygen of, or in excess of, 135 tonnes.

7. Any premises at which there are stored, or there are facilities provided for the storage of, quantities of chlorine of, or in excess of, 50 tonnes except where the chlorine is kept solely for the purpose of water purification.

8. Any premises at which artificial fertilizers are manufactured and at which there are stored, or there are facilities provided for the storage of, quantities of ammonia of, or in excess of, 250 tonnes.

9. Any premises at which there are in process, manufacture, use or storage at any one time, or there are facilities provided for such processing, manufacture, use or storage of, quantities of any of the materials listed below in, or in excess of, the quantities specified —

Phosgene	5 tonnes
Ethylene oxide	20 tonnes
Carbon disulphide	50 tonnes
Acrylonitrile	50 tonnes
Hydrogen cyanide	50 tonnes
Ethylene	100 tonnes
Propylene	100 tonnes
Any highly flammable liquid not otherwise specified	4000 tonnes

10. Explosives factories or magazines which are required to be licensed under the Explosives Act 1875.

11. Any building on the surface at any mine within the meaning of the Mines and Quarries Act 1954.

12. Any premises in which there is comprised —
(a) any undertaking on a site for which a licence is required in accordance with Section 1 of the Nuclear Installations Act 1965 or for which a permit is

required in accordance with Section 2 of that Act; or
(b) any undertaking which would, except for the fact that it is carried on by the United Kingdom Atomic Energy Authority, or by, or on behalf of, the Crown, be required to have a licence or permit in accordance with the provisions mentioned in sub-paragraph (a) above.

13. Any premises containing any machine or apparatus in which charged particles can be accelerated by the equivalent of a voltage of not less than 50 megavolts except where the premises are used as a hospital.

14. Any premises at which there are in process, manufacture, use or storage at any one time, or there are facilities provided for such processing, manufacture, use or storage of, quantities of unsealed radioactive substances classified according to Schedule 3 of the Ionising Radiations (Unsealed Radioactive Substances) Regulations 1968 in, or in excess of the quantities specified —

Class I radionuclides	—	10 curies
Class II and III radionuclides	—	100 curies
Class IV radionuclides	—	1000 curies

15. Any building, or part of a building, which either —
(a) is constructed for temporary occupation for the purposes of building operations or works of engineering construction; or
(b) is in existence at the first commencement there of any further such operations or works
and which is used for any process or work ancillary to any such operations or works.

D. Building Act 1984

Section 1. 'Power to make building regulations'

(1) The Secretary of State may, for any of the purposes of:—
   (a) securing the health, safety, welfare and convenience of persons in or about buildings and of others who may be affected by buildings or matters connected with buildings,
   (b) furthering the conservation of fuel and power, and
   (c) preventing waste, undue consumption, misuse or contamination of water,
   make regulations with respect to the design and construction of buildings and the provision of services, fittings and equipment in or in connection with buildings.

(2) Regulations made under subsection (1) above are known as building regulations.

(3) Schedule 1 to this Act has effect with respect to the matters as to which building regulations may provide.

(4) The power to make building regulations is exercisable by statutory instrument, which is subject to annulment in pursuance of a resolution of either House of Parliament.

Section 4. 'Exemption of educational buildings and buildings of statutory undertakers'

(1) (a) state schools.
    (b) statutory undertakers, the UK Atomic Energy Authority, the British Airports Authority and the Civil Aviation Authority.

Section 5. 'Exemption of public bodies from procedural requirements of building regulations'

Section 6. 'Approval of documents for purposes of building regulations'

Section 24. 'Provision of exits etc'

Power of local authority* to reject plans deposited in accordance with building regulations where the means of ingress and egress and passages or gangways are not deemed satisfactory — regard being had to the purposes for which the building is intended to be, or is, used and the number of persons likely to resort to it at any one time. Applies to:—
(a) a theatre, and a hall or other building that is used as a place of public resort,
(b) a restaurant, shop, store or warehouse to which members of the public are admitted and in which more than twenty persons are employed,
(c) a club required to be registered under the Licensing Act 1964,
(d) a school not exempted from the operation of building regulations, and
(e) a church, chapel or other place of public worship.

But not if requirements for means of escape are imposed under building regulations; or to a private house to which members of the public are admitted occasionally or exceptionally; or to a building used as a church, chapel or other place of public worship immediately before the date on which s.36 of the Public Health Acts Amendment Act 1890 came into operation (or a corresponding provision in a local Act) or before 1.10.1937 where such provisions had not come into operation.

Section 71. 'Entrances, exits etc to be required in certain cases'

Applies to existing buildings to which Section 24 above applies.

Section 72. 'Means of escape from fire'

Power of local authority* to require the provision of satisfactory means of escape from floors over 20 ft above the street or ground level in new and existing

---

*After consulting the fire authority.

buildings. Applies to:—

Any building which exceeds two storeys in height and in which the floor of any upper storey is more than twenty feet above the surface of the street or ground on any side of the building and that:—
(a) is let in flats or tenement dwellings,
(b) is used as an inn, hotel, boarding-house, hospital, nursing home, boarding-school, children's home or similar institution, or
(c) is used as a restaurant, shop, store or warehouse and has on an upper floor sleeping accommodation for persons employed on the premises.

But not if requirements for means of escape are imposed under building regulations.

Schedule 1. Building Regulations

(7) Without prejudice to the generality of section 1(1) of this Act, building regulations may:—
    (a) for any of the purposes mentioned in section 1(1) of this Act, make provision with respect to any of the following matters:—
        (iv) fire precautions, including —
            (a) structural measures to resist the outbreak and spread of fire and to mitigate its effects,
            (b) services, fittings and equipment designed to mitigate the effects of fire or to facilitate fire-fighting,
            (c) means of escape in case of fire and means for securing that such means of escape can be safely and effectively used at all material times.

E. The Building Regulations 1991

Schedule 1, Part B, Fire Safety

B1. *Means of escape**

The building shall be designed and constructed so that there are means of escape in case of fire from the building to a place of safety outside the building capable of being safely and effectively used at all material times.

B2. *Internal fire spread (linings)*

(1) To inhibit the spread of fire within the building the internal linings shall —

    (a) resist the spread of flame over their surfaces; and
    (b) have, if ignited, a rate of heat release which is reasonable in the circumstances.

---

* This requirement does not apply to any prison provided under section 33 of The Prisons Act 1952.

(2) In this paragraph "internal linings" mean the materials lining any partition, wall, ceiling or other internal structure.

B3. *Internal fire spread (structure)*

(1) The building shall be designed and constructed so that, in the event of fire, its stability will be maintained for a reasonable period.

(2) A wall common to two or more buildings shall be designed and constructed so that it resists the spread of fire between those buildings. For the purposes of this sub-paragraph a house in a terrace and a semi-detached house are each to be treated as a separate building.

(3) To inhibit the spread of fire within the building, it shall be sub-divided with fire-resisting construction to an extent appropriate to the size and intended use of the building.

(4) The building shall be designed and constructed so that the unseen spread of fire and smoke within concealed spaces in its structure and fabric is inhibited.

B4. *External fire spread*

(1) The external walls of the building shall resist the spread of fire over the walls and from one building to another, having regard to the height, use and position of the building.

(2) The roof of the building shall resist the spread of fire over the roof and from one building to another, having regard to the use and position of the building.

B5. *Access and facilities for the fire service*

(1) The building shall be designed and constructed so as to provide facilities to assist fire fighters in the protection of life.

(2) Provision shall be made within the site of the building to enable fire appliances to gain access to the building.

Appendix 3: Standard fire tests

BS 476: Fire tests on building materials and structures
Part 3 (1958) External fire exposure roof tests
This standard contains two methods of test which enable the measurement of
(a) the capacity of a representative section of a roof to resist penetration by fire when the external surface is exposed to radiation and flame; and
(b) the distance of the spread of flame on the outer surface of the roof covering under certain conditions.

In each test three representative sections of the roof (840 mm square) are exposed to heat radiated from gas fired panels (Figure 24); a pilot flame representing a flying burning brand is applied to the hot roof. A preliminary screening test with the pilot flame is carried out on one sample without the radiation.

Figure 24 External fire exposure roof test (BS 476: Part 3: 1958) (specimen arrowed)

The radiation to which the specimens are subjected in the penetration test, in terms of practical fires, would represent the radiation* on a roof 7.5 m (25 ft) above ground level at a distance of 13.5 m (45 ft) from a burning building which has a facade of 15 m × 15 m (50 ft × 50 ft) with 50 per cent of the elevation as windows. For the assessment of flame spread, the radiation* varies between that which would be experienced at equivalent distances of 18 m (60 ft) to 36 m (120 ft) (Figure 25). These levels are low in comparison with the gradient specified in BS 476: Part 7 and very much lower than the heating conditions used in BS 476: Part 20.

Provision is also made in the test for penetration to simulate the effect of wind of approximately 15 mph by applying suction to the lower side of the specimen during the test.

After testing, constructions are designated by two letters:
    First letter — time to penetration
    Second letter — distance of spread of flame along their external surface

The highest classification is AA and the lowest DD, the designations being as indicated in Table 14. However, the tests cannot be used for some roof constructions. For example, some types of plastics materials may not ignite but will fall away prematurely.

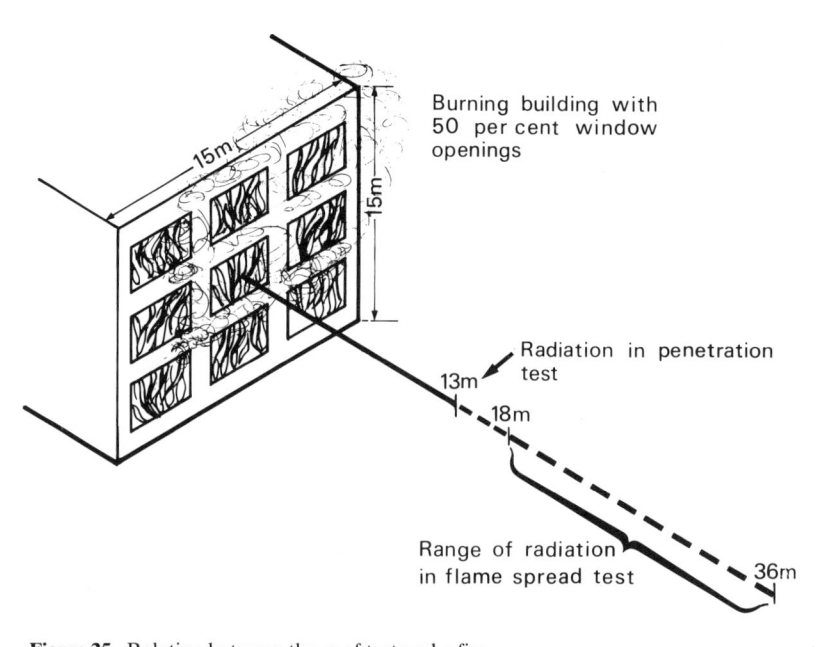

Figure 25 Relation between the roof test and a fire

---

* Strictly 'irradiance'

Table 14 Classification on roof test (1958)

	Time to penetration			
Flame spread	Nil in 1 hour	Nil in ½ hour	Up to ½ hour	Preliminary test
No flame spread	AA	BA	CA	DA
Flame spread up to 21 in	AB	BB	CB	DB
Flame spread over 21 in	AC	BC	CC	DC
Sustained ignition* from pilot flame	AD	BD	CD	DD

*Continues to burn for 5 minutes after the withdrawal of the test flame or spread of flame more than 15 in across region of burning.

Roof designations are preceded by the letters EXT.F or EXT.S according to whether the roof specimen represented a flat or sloping construction (ie tested horizontally or at an angle of 45°). A suffix X is added where dripping occurs from the underside of the specimen, or any mechanical failure or any holes appear in it during the test.

Part 3 (1975) External fire exposure roof test
In this revision of the above standard, a number of modifications were made. The most notable was the decision that the performance of the specimen is not expressed in terms of definite designations and is replaced by the actual performance data, viz a P60 result means that the specimen passed the preliminary test and that fire penetration did not occur in less than 1 hour.

Other modifications include: replacement of the spread of flame test in the earlier edition by a measurement of the extent of surface ignition which is made at the same time as the penetration test; the requirement for suction under the roof specimen has been omitted; the duration of the test may be extended beyond 60 minutes to a maximum of 90 minutes; the sample size has been increased to 1.5 m × 1.2 m; and the number of specimens has been reduced to four.

Part 4 (1970) Non-combustibility test for materials
This test determines whether building materials are non-combustible within the meaning of the definition in this standard. The apparatus (Figure 26) is a cylindrical furnace preheated to 750°C into which a 40 mm × 40 mm × 50 mm sample is exposed for 20 minutes. Three specimens are tested.

Determination of non-combustibility is made where during the test none of the specimens either —
(i) causes the temperature reading from either of the two thermocouples (one for the furnace and one at the centre of the specimen) to rise by 50°C or more above the initial furnace temperature; or
(ii) is observed to flame continuously for 10 seconds or more inside the furnace.

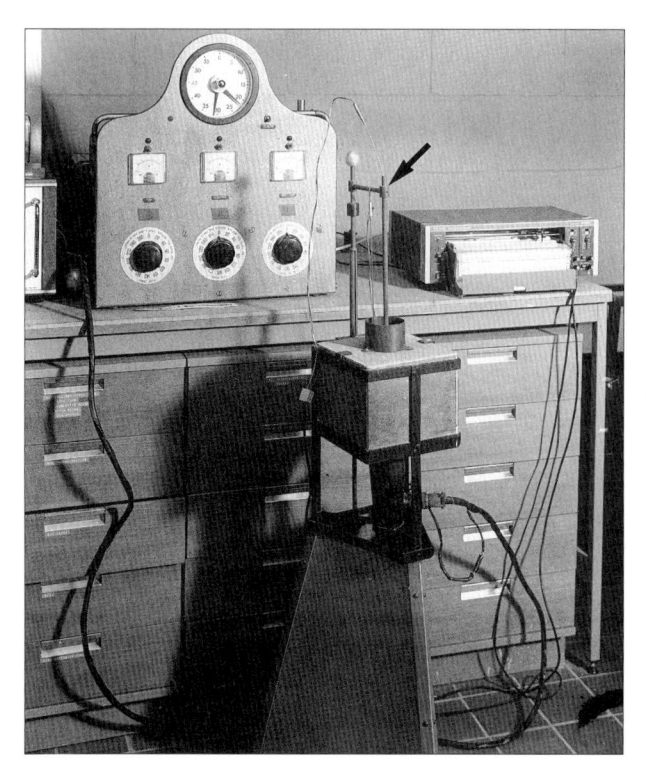

Figure 26 Non-combustibility test apparatus (BS 476: Part 4)
Specimen is suspended from insertion rod (arrowed)

Part 5 (1979) Method of test for ignitability *(withdrawn)*
In this test (Figure 27) a 225 mm × 225 mm sample (of the normal thickness of the product), held vertically, is subjected to a small flame for 10 seconds. The test classifies materials as either X or P (where X is the lower performance). Materials which achieve P, however, cannot necessarily be considered to be of low risk, as compliance only implies that the material will not ignite when exposed (for example) to a match flame under the conditions of this test. Three specimens are tested.

Part 6 (1989) Method of test for fire propagation for products
This test provides a means of comparing the contribution of combustible building materials to the growth of a fire by providing a measure of the rate of heat evolution of a 225 mm × 225 mm sample, up to 50 mm thick, exposed in a small combustion chamber (Figure 28) for 20 minutes to a specified heating regime of continuously increasing severity. The gas jets are ignited at the start of the test, with the electric radiant bars being added after $2^3/4$ minutes.

The performance is expressed as a numerical index from 0 to 100 or more and is based on the readings of a thermocouple inside the cowl being compared with those on a calibration curve obtained using asbestos board samples. Low values indicate a low rate of heat release. Three to five specimens are tested.

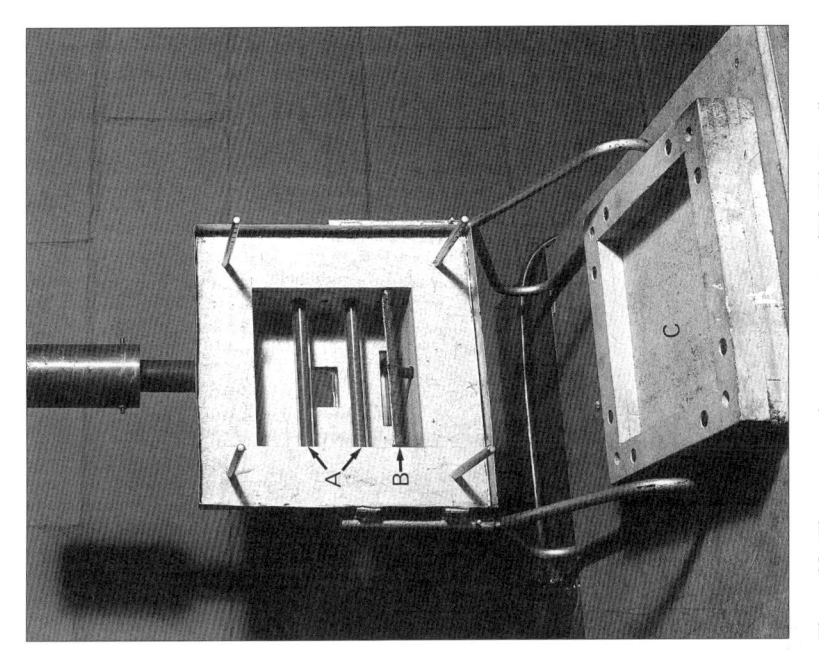

Figure 28 Fire propagation test apparatus (BS 476: Part 6)
A : electric radiant bars   B : gas jets
C : specimen is placed in recess bolted against the face of the furnace

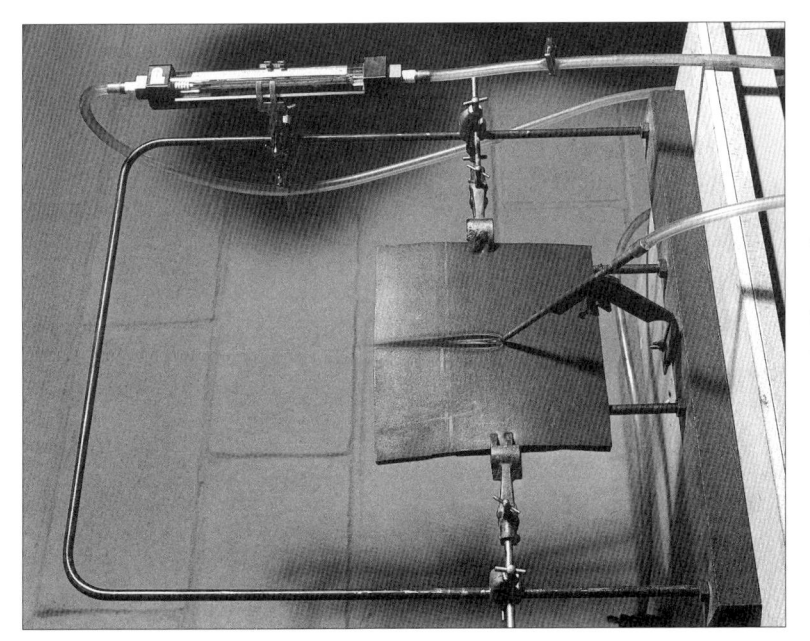

Figure 27 Ignitability test (BS 476: Part 5)

Index of performance $I = i_1 + i_2 + i_3$ where sub-index i_1 is derived from the first 3 minutes of test, i_2 from the following 7 minutes, and i_3 from the final 10 minutes. A high index i_1 indicates an initial rapid ignition and heat release. If four or five specimens have to be tested to achieve three valid results, a suffix 'R' is added to indicate variability in test results.

Part 7 (1987) Method for classification of the surface spread of flame of products
This test is used to determine the tendency of essentially flat materials to support the spread of flame across their surfaces and specifies a method of classification appropriate to wall and ceiling linings. Specimens are exposed to a 900 mm square radiant panel (Figure 29) which is run at a temperature of 800°C – 1000°C with the intensity of radiation on the specimen varying from 32.5 kW/m^2 – 5.0 kW/m^2. For the first minute of test a pilot flame is used to ignite flammable volatiles at the hot end of the specimen. Six to nine specimens 885 mm × 270 mm and up to 50 mm thick are tested.

Figure 29 Surface spread of flame test (BS 476: Part 7)
Specimen located in holder arrowed

The extent of flame spread after 1½ minutes and at the end of the ten minute test is used to classify* products as shown in Figure 30 (with a permitted tolerance for one specimen in the sample) — Class 1 representing the best performance. If seven, eight or nine specimens have to be tested to achieve the necessary six valid results, a suffix 'R' is added to indicate variability in test results. A suffix 'Y' is added if softening and/or other behaviour occurs which may affect the flame spread, and a prefix 'D' for those products tested in a modified form (eg corrugated sheeting tested flat).

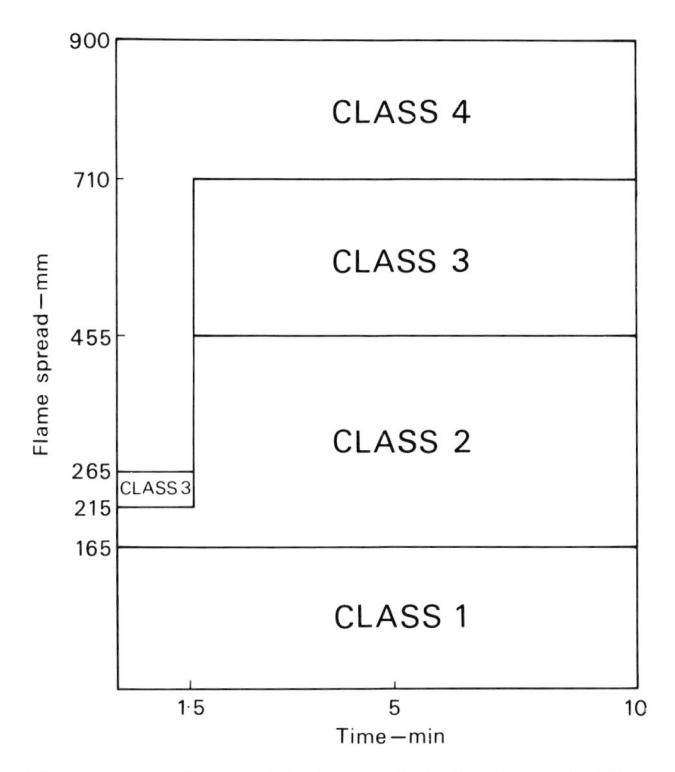

Figure 30 An indication of classification limits for the spread of flame test (based on Figure 2 of BS 476: Part 7: 1971)

Part 11 (1982) Method for assessing the heat emission from building materials

This standard is a development of BS 476: Part 4. However, instead of classifying materials by reference to an arbitrary level of combustibility, results are given in terms of the behaviour of the material in the test. Each specimen (45 mm in diameter, with a height not exceeding 50 mm) is exposed for up to 120 minutes in the preheated furnace (Figure 26) and note is taken of any rise in temperature of both the furnace and specimen thermocouples in excess of the final equilibrium temperature, and of any continuous flaming of a least 5 seconds duration. From these results are calculated: the mean furnace temperature rise (°C), the mean

---

*Class 0 is not a BSI method of classification. It is a term defined in connection with Building Regulations which makes use of BSI performance ratings (see Chapter 5).

specimen temperature rise (°C), and the mean duration of sustained flaming (seconds). Five specimens are normally tested.

Part 13 (1987) Method of measuring the ignitability of products subjected to thermal irradiance

This test (Figure 31), which is identical to ISO 5657: 1986, examines the ability of products to become ignited when subjected to radiation in the presence of means for pilot ignition, eg flame contact, embers, sparks or falling burning material. It is intended to provide information which can be used in the evaluation of wall and ceiling linings, flooring systems, external cladding and duct insulation materials.

The upper surface of each specimen (165 mm square and not exceeding 70 mm in thickness) is exposed for up to 15 minutes to selected levels of constant radiation within the range 10 kW/m^2 – 50 kW/m^2. A pilot flame is applied at 4 second intervals to a position 10 mm above the centre of the specimen to ignite any volatile gases given off, and the time to any sustained surface ignition (in seconds) is reported. Five specimens are tested at each level of irradiance selected and for each different exposed surface.

A method of test for ignitability of products by direct flame impingement forms the subject of Part 12.

Figure 31 Ignitabililty test (BS 476: Part 13) (specimen arrowed)

Parts 20 – 23 (1987) Methods for determination of the fire resistance of elements of construction
These standards constitute a complete revision of BS 476: Part 8. In conjunction with Part 20, methods of test are provided for — Beams, columns, floors, flat roofs and walls (Part 21); Partitions, doorsets and vertical shutter assemblies, ceiling membranes and glazed elements (Part 22); and Suspended ceilings protecting steel beams and for intumescent seals for use in conjunction with single-acting latched timber fire-resisting door assemblies (Part 23). It is expected that Part 23 will be extended by tests for the contribution made by other components.

Part 20 specifies standard heating conditions based on a temperature/time curve (Figure 32) which furnaces are required to follow, the temperature at defined locations close to the exposed face of the specimen under test rising to 821°C after 30 minutes and 1133°C after 4 hours. The specimen to be tested should be either full-size or, where the element exceeds the size that can be accommodated by the furnace, it must have the following minimum dimensions:

non-separating elements: vertical 3 m high
                                     horizontal 4 m span
separating elements: vertical 3 m high × 3 m wide
                                     horizontal 4 m high × 3m wide

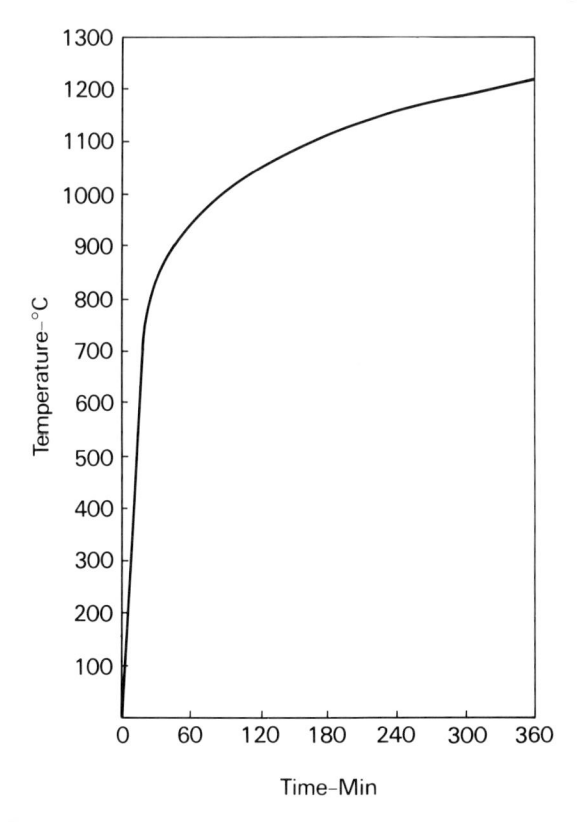

Figure 32   Standard temperature/time curve (BS 476: Part 20: 1987)

Specimens are normally heated to simulate their exposure in a fire, eg walls from one side, floors from beneath and columns from all sides. Only one specimen is tested. Figures 33, 34 and 35 show a wall, column and floor furnace respectively.

Elements of building construction are required to satisfy various criteria according to their designed function in the event of fire. These are:

'loadbearing capacity' — the ability of a loadbearing element to support its test load without excessive deflection;

'integrity' — the ability of a separating element to resist collapse, the formation of holes, gaps or fissures through which flames and/or hot gases could pass, and the occurrence of sustained flaming on the unexposed face;

'insulation' — the ability of a separating element to resist an excessive rise in temperature on its unexposed face.

The criterion of 'loadbearing capacity' is applied only to loadbearing elements (under Part 8, 'stability' also applied to non-loadbearing elements). For floors, flat roofs and beams, allowable vertical deflection is limited to 1/20 clear span.

Loss of 'integrity' in the context of the formation of holes, gaps or fissures is judged by ignition of a cotton fibre pad. Where this test is not suitable, failure is deemed to have occurred if either a 25 mm diameter gauge can penetrate into the furnace through a gap at any point, or a 6 mm diameter gauge can penetrate into the furnace through an opening and can be moved for a distance of at least 150 mm.

Loss of 'insulation' occurs when the temperature on the unexposed face (the side of the specimen remote from the furnace) increases by more than 140°C (mean) or by more than 180°C at any point. Loss of 'integrity' also constitutes loss of 'insulation'.

Columns and beams have to satisfy only the criterion of 'loadbearing capacity'; glazed elements are normally required to satisfy only 'integrity' (requirements made in connection with regulations etc, normally permit doors to be similarly judged since it is unlikely that combustible materials would be stored against them); and floors and walls have to satisfy all three criteria.

Although the performance is expressed as fire resistance in minutes for each of the relevant criteria, requirements made in connection with regulations, etc frequently refer to 1/2 h, 1 h, 1 1/2 h, 2 h, 3 h and 4 h periods. For all tests, the furnace is maintained at a positive pressure relative to the laboratory; whereas under Part 8 this was required only for specimens tested for 'integrity'.

Part 24 (1987) Method for determination of the fire resistance of ventilation ducts
This test (identical to ISO 6944:1985) measures the ability of vertical and horizontal ductwork to resist the spread of fire from one fire compartment to another without the aid of fire dampers.

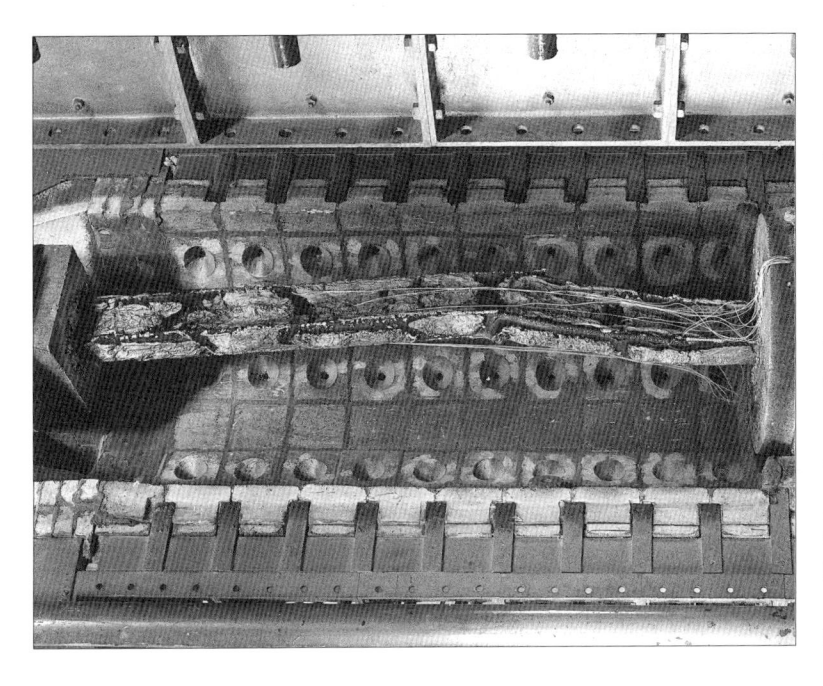

Figure 34 Column furnace showing a steel column after test

Figure 33 Lighting wall furnace prior to test of loadbearing specimen

Figure 35 Floor furnace showing a timber floor collapsing

The specimen to be tested should be full-size or, where this is not possible, have the following minimum dimensions:

	horizontal ductwork	vertical ductwork
portion in the furnace	length 3 m	length 2 m
portion outside the furnace	length 2.5 m	length 2 m

Two sections are tested together: one (A) which does not have an opening within the furnace and the other (B) which has. A fan is fitted on the end outside the furnace of section 'B' (to extract the furnace gases at an air velocity of at least 3 m/s) and of horizontal section 'A' (to produce an underpressure of 300 Pa).

Results are given in terms of performance (in minutes) for each of the criteria of stability*, integrity† and insulation. (The reference to 'stability' is due to the standard referring to ISO 834.)

Additional criteria may be needed in respect of the integrity of smoke outlet ductwork and of the insulation performance of ductwork lined on the inside with combustible material (or which in practice may accumulate combustible deposits on the inside face, eg kitchen extract ductwork).

---

*Collapse outside the furnace; and inside for 'A'.
† Measured for both "fan on" and "fan off" situations.

Section 31.1 (1983) Method for measuring smoke penetration through doorsets and shutter assemblies (ambient temperature conditions)

This test (based on ISO 5925-1) determines the extent of smoke penetration as represented by measurement of air leakage rate with the assembly fitted to form one side of a chamber (Figure 36). The rate of air leakage is established by measuring the airflow for increasing pressure differentials* between the two faces of the specimen, up to the maximum pressure differential for which information is required. Results are given of the adjusted rate of air leakage (m^3/h) for each pressure differential and each face of the specimen tested. One or two full-sized specimens are tested depending on whether provision exists for reversing either the direction of airflow through the test chamber, or the specimen in its surround.

Medium temperature conditions will be dealt with under Section 31.2 and high temperature conditions under Section 31.3

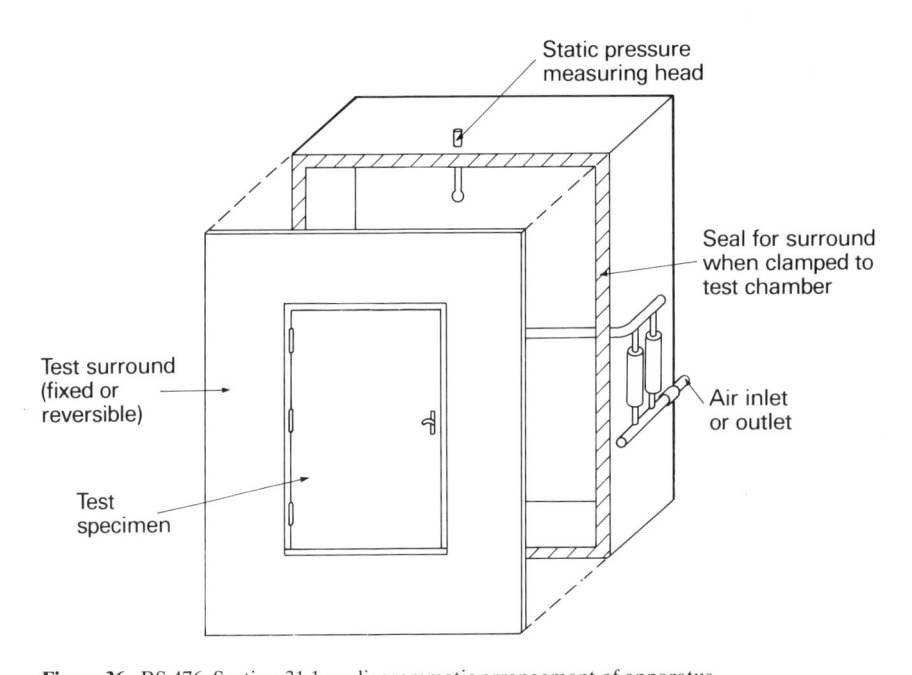

Figure 36 BS 476: Section 31.1 — diagrammatic arrangement of apparatus

---

* 5, 10, 20, 30, 50, 70 and 100 Pa.

BS 2782: Methods of testing plastics

This standard has separate parts for measuring various properties of plastics materials as a means of quality control. Those methods of test which have been adopted for use in connection with regulations for determining acceptable performance with respect to specific characteristics of fire behaviour are included below. However, these small-scale tests are not for use as a means of assessing the potential fire hazard of a material in use.

Method 120A (1976) Determination of the Vicat softening temperature of thermoplastics

This is the basic procedure for determining the Vicat softening temperature (VST) of thermoplastics under compression. A 10 mm square (min) × 2.5 mm – 6.5 mm thick sample is heated in a bath of an appropriate liquid (eg glycerol for testing polystyrene and rigid PVC). A force of 10 N is applied to the specimen and note is taken of the temperature at which the indenting tip penetrates 1 mm into the specimen (Figure 37). The VST is taken as the arithmetic mean of a pair of satisfactory results. This standard was revised in 1990.

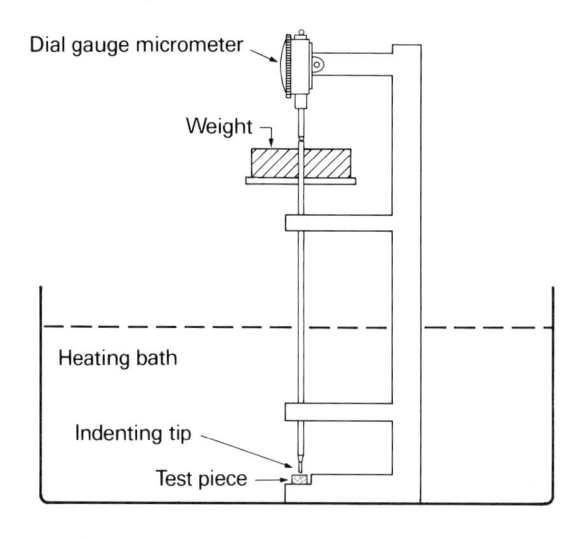

Figure 37
BS 2782 Method 120A
— diagrammatic arrangement of apparatus

Method 508A (1970) Rate of burning (laboratory method)

At least three specimens of the material are tested — 150 mm long × 13 mm wide × 1.5 mm thick, with two marks 100 mm apart (the first 25 mm from one end). Each specimen is held at 45° (the long axis horizontal) with the end with the 25 mm mark exposed for 10 seconds to a bunsen flame 13 mm – 19 mm long (Figure 38). After the bunsen flame is removed the extent and rate of flame spread is noted.

For the purposes of Building Regulations, 1.5 mm – 3 mm thick specimens are required to be tested in connection with control on the use of certain thermoplastic rooflights and lighting diffusers.

This test has been superseded by Method 140A: 1992 (Method A).

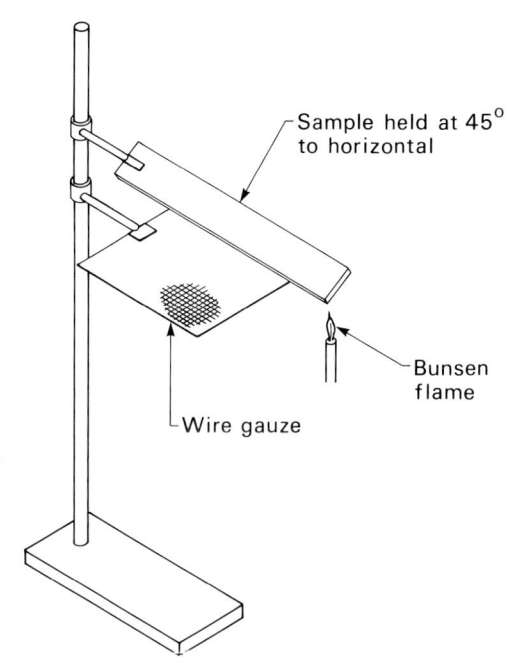

Figure 38 BS 2782 Method 508A — diagrammatic arrangement of specimen and apparatus

Method 140D (1980) Flammability of a test piece 550 mm × 35 mm of thin PVC sheeting (laboratory method)

The strip of sheeting is held over a semi-circular frame (Figure 39) — one end of the specimen being exposed to 0.1 millilitre of burning alcohol. The test result is reported as the distance (mm) over which the strip has burned under these conditions. Six specimens are tested.

This test is basically the same as Method 508C (1970) which it supersedes.

Method 140E (1982) Flammability of a small inclined test piece exposed to an alcohol flame (laboratory method)

A 150 mm square × 50 mm (max) thick sample is held at 45° above 0.3 millilitre of burning alcohol such that the flame impinges centrally on the underside 25 mm above the alcohol container (Figure 40). For each test piece, the duration of glowing or flaming after the alcohol has burnt out is reported, together with a note of any fallen burning material and the length and percentage of the area of the underside which is charred or scorched. Three specimens are tested.

This test is basically the same as Method 508D (1970) which it supersedes.

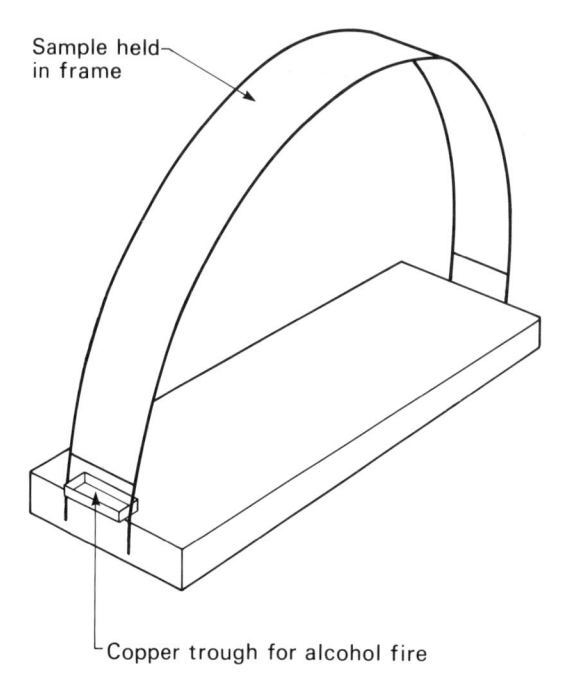

Figure 39 BS 2782 Method 140D — diagrammatic arrangement of apparatus

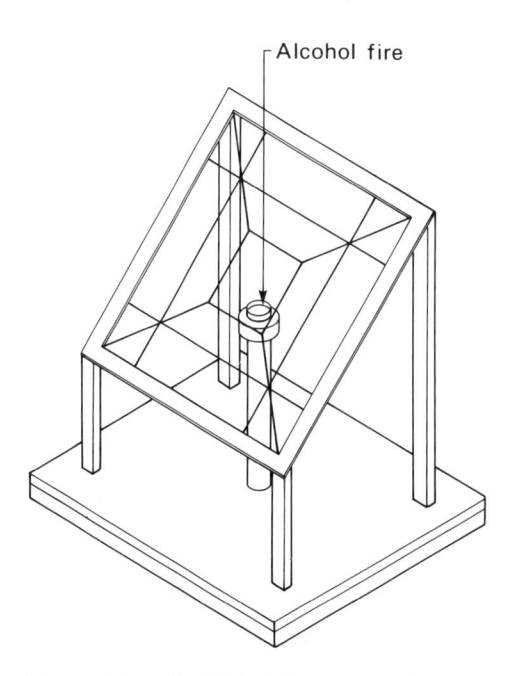

Figure 40 BS 2782 Method 140E — general arrangement of apparatus

Appendix 4: Reading list and other relevant sources of information

General
- The Fire Research Station, Borehamwood, Herts, WD6 2BL.
- The Fire Protection Association, 140 Aldersgate Street, London, EC1A 4HX.
- The Loss Prevention Council, 140 Aldersgate Street, London, EC1A 4HX.
- The Timber Research and Development Association, High Wycombe, Bucks, HP14 4ND.

Books
- **Butcher E G and Parnell A C.** *Smoke control in fire safety design.* London, E & F N Spon, 1979.
- **Langdon-Thomas G J.** *Fire Safety in Buildings. Principles and Practice.* London, A & C Black Ltd, 1972.
- **Malhotra H L.** *Design of fire-resisting structures.* Glasgow, Surrey University Press, 1982.
- **Marchant E W.** *A complete guide to fire and buildings.* Lancaster, Medical and Technical Publishing Co Ltd, 1972.
- **Read R E H.** *British statutes relating to fire 1425–1963.* Building Research Establishment Report. Garston, BRE, 1986.
- **Read R E H.** *British fire legislation on means of escape 1774–1974.* Building Research Establishment Report. Garston, BRE, 1986.
- **Home Office.** *Manual of Firemanship*, Book 8, Building construction and structural fire protection. London, HMSO, 1992.
- **Home Office.** *Manual of Firemanship*, Book 9, Fire protection of buildings. London, HMSO, 1991.

Statutory Instruments*
England and Wales
- FIRE PRECAUTIONS. The Fire Precautions (Hotels and Boarding Houses) Order 1972. (SI 1972 No 238).
- HEALTH AND SAFETY. Fire Certificates (Special Premises) Regulations 1976. (SI 1976 No 2003).
- HOUSING, ENGLAND AND WALES. The Housing (Means of Escape from Fire in Houses in Multiple Occupation) Order 1981. (SI 1981 No 1576).

---

*Available from HMSO.

- BUILDING AND BUILDINGS. The Building (Inner London) Regulations 1985. (SI 1985 No 1936).
- BUILDING AND BUILDINGS. The Building (Inner London) Regulations 1987. (SI 1987 No 798).
- FIRE PRECAUTIONS. SPORTS GROUNDS AND SPORTING EVENTS. The Fire Safety and Safety of Places of Sport Act 1987 (Commencement No 1) Order 1987. (SI 1987 No 1762 (C.54)).
- FIRE PRECAUTIONS. The Fire Safety and Safety of Places of Sport Act 1987 (Commencement No 5) Order 1989. (SI 1989 No 75 (C.3)).
- FIRE PRECAUTIONS. The Fire Precautions (Factories, Offices, Shops and Railway Premises) Order 1989. (SI 1989 No 76).
- FIRE PRECAUTIONS. The Fire Precautions (Sub-surface Railway Stations) Regulations 1989. (SI 1989 No 1401).
- BUILDING AND BUILDINGS. The Construction Products Regulations 1991. (SI 1991 No 1620).
- BUILDING AND BUILDINGS. The Building Regulations 1991. (SI 1991 No 2768).
- FIRE PRECAUTIONS. The Fire Safety and Safety of Places of Sport Act 1987 (Commencement No 7) Order 1993. (SI 1993 No 1411 (C.26)).

Scotland
- FIRE PRECAUTIONS. The Fire Precautions (Hotels and Boarding Houses) (Scotland) Order 1972. (SI 1972 No 382 (S.26)).
- HEALTH AND SAFETY. Fire Certificates (Special Premises) Regulations 1976. (SI 1976 No. 2003).
- FIRE PRECAUTIONS. SPORTS GROUNDS AND SPORTING EVENTS. The Fire Safety and Safety of Places of Sport Act 1987 (Commencement No 1) Order 1987. (SI 1987 No 1762 (C.54)).
- FIRE PRECAUTIONS. The Fire Safety and Safety of Places of Sport Act 1987 (Commencement No 5) Order 1989. (SI 1989 No 75 (C.3)).
- FIRE PRECAUTIONS. The Fire Precautions (Factories, Offices, Shops and Railway Premises) Order 1989. (SI 1989 No 76).
- FIRE PRECAUTIONS. The Fire Precautions (Sub-surface Railway Stations) Regulations 1989. (SI 1989 No 1401).
- BUILDING AND BUILDINGS. The Building Standards (Scotland) Regulations 1990. (SI 1990 No 2179 (S.187)).
- FIRE PRECAUTIONS. The Fire Precautions (Sub-surface Railway Stations) (Amendment) Regulations 1991. (SI 1991 No 259 (S.21)).
- BUILDING AND BUILDINGS. The Construction Products Regulations 1991. (SI 1991 No 1620).
- FIRE PRECAUTIONS. The Fire Safety and Safety of Places of Sport Act 1987 (Commencement No 7) Order 1993. (SI 1993 No 1411 (C.26)).

Northern Ireland
- [†] NORTHERN IRELAND. The Fire Services (Northern Ireland) Order 1984. (SI 1984 No 1821 (NI.11)).

[†] In course of revision.

- FIRE SERVICES. The Fire Services (Leisure Premises) Order (Northern Ireland) 1985. (SRNI 1985 No 137).
- FIRE SERVICES. The Fire Services (Hotels and Boarding Houses) Order (Northern Ireland) 1985. (SRNI 1985 No 138).
- FIRE SERVICES. The Fire Services (Non-Certificated Factory, Office and Shop Premises) Regulations (Northern Ireland) 1986. (SRNI 1986 No 352).
- FIRE SERVICES. The Fire Services (Factory, Office and Shop Premises) Order (Northern Ireland) 1986. (SRNI 1986 No 355).
- FIRE SERVICES. The Fire Services (Betting, Gaming and Amusement Premises) Order (Northern Ireland) 1987. (SRNI 1987 No 334).
- BUILDING REGULATIONS. The Building Regulations (Northern Ireland) 1990. (SRNI 1990 No 59).
- HEALTH AND SAFETY. Fire Certificates (Special Premises) Regulations (Northern Ireland) 1991. (SRNI 1991 No 446).

British Standards Institution (BSI)[§]

- BS 476. Fire tests on building materials and structures
    - Part 1. 1953. Fire tests on building materials and structures (*withdrawn*)
    - Part 2. 1955. Flammability test for thin flexible materials (*withdrawn*)
    - Part 3. 1975. External fire exposure roof test
    - Part 4. 1970. Non-combustibility test for materials
    - Part 5. 1979. Method of test for ignitability *(withdrawn)*
    - Part 6. 1989. Method of test for fire propagation for products
    - Part 7. 1987. Method for classification of the surface spread of flame of products
    - Part 8. 1972. Test methods and criteria for the fire resistance of elements of building construction (*withdrawn*)
    - Part 10. 1983. Guide to the principles and application of fire testing
    - Part 11. 1982. Method for assessing the heat emission from building materials
    - Part 12. 1991. Method of test for ignitability of products by direct flame impingement
    - Part 13. 1987. Method of measuring the ignitability of products subjected to thermal irradiance (ISO 5657)
    - Part 20. 1987. Method for determination of the fire resistance of elements of construction (general principles)
    - Part 21. 1987. Methods for determination of the fire resistance of loadbearing elements of construction
    - Part 22. 1987. Methods for determination of the fire resistance of non-loadbearing elements of construction
    - Part 23. 1987. Methods for determination of the contribution of components to the fire resistance of a structure
    - Part 24. 1987. Method for determination of the fire resistance of ventilation ducts (ISO 6944)

[§] Available from BSI Sales Department, Linford Wood, Milton Keynes, MK14 6LE.

       Section 31.1. 1983. Methods for measuring smoke penetration through doorsets and shutter assemblies. Method of measurement under ambient temperature conditions
       Part 32. 1989. Guide to full scale fire tests within buildings
- BS 1635. 1990. Recommendations for graphic symbols and abbreviations for fire protection drawings
- BS 2782. 1970. Methods of testing plastics
- BS 2782. Methods of testing plastics. Part 1. Thermal properties
  Methods 120A, 120B, 120D and 120E. 1990. Determination of Vicat softening temperature of thermoplastics (ISO 306)
  Method 140A. 1992. Determination of the burning behaviour of horizontal and vertical specimens in contact with a small-flame ignition source (ISO 1210:1992)
  Method 140D. 1980. Flammability of a test piece 550 mm × 35 mm of thin polyvinyl chloride sheeting (laboratory method)
  Method 140E. 1982. Flammability of a small inclined test piece exposed to an alcohol flame (laboratory method)
- BS 4422. Glossary of terms associated with fire
      Part 1. 1987. General terms and phenomena of fire (ISO 8421–1)
      Part 2. 1990. Structural fire protection
      Part 3. 1990. Fire detection and alarm (ISO 8421–3)
  [†] Part 4. 1975. Fire protection equipment
      Part 5. 1989. Smoke control (ISO 8421–5)
      Part 6. 1988. Evacuation and means of escape (ISO 8421–6)
      Part 7. 1988. Explosion detection and suppression means (ISO 8421–7)
  [‡] Part 8. Terms specific to firefighting and rescue services
      Part 9. 1990. Marine terms
- BS 5266. Part 1. 1988. Emergency lighting. Code of practice for the emergency lighting of premises other than cinemas and certain other specified premises used for entertainment
- BS 5268. Structural use of timber. Part 4. Fire resistance of timber structures
      Section 4.1. 1978. Recommendations for calculating fire resistance of timber members
      Section 4.2. 1990. Recommendations for calculating fire resistance of timber stud walls and joisted floor constructions
- BS 5306. Fire extinguishing installations and equipment on premises
      Part 0. 1986. Guide for the selection of installed systems and other fire equipment
      Part 1. 1976. Hydrant systems, hose reels and foam inlets
      Part 2. 1990. Specification for sprinkler systems
      Part 3. 1985. Code of practice for selection, installation and maintenance of portable fire extinguishers
      Part 4. 1986. Specification for carbon dioxide systems
      Part 5. Halon systems
        Section 5.1. 1982. Halon 1301 total flooding systems

[†] In course of revision.
[‡] In course of preparation.

  Section 5.2. 1984. Halon 1211 total flooding systems
  Part 6. Foam systems
   Section 6.1. 1988. Specification for low expansion foam systems
   Section 6.2. 1989. Specification for medium and high expansion foam systems
  Part 7. 1988. Specification for powder systems
- BS 5378. Part 1. 1980. Safety signs and colours. Specification for colour and design
- BS 5395. Stairs, ladders and walkways
  Part 1. 1977. Code of practice for the design of straight stairs
  Part 2. 1984. Code of practice for the design of helical and spiral stairs
  Part 3. 1985. Code of practice for the design of industrial type stairs, permanent ladders and walkways
- BS 5422. 1990. Method for specifying thermal insulating materials on pipes, ductwork and equipment (in the temperature range −40°C to +700°C)
- BS 5438. 1989. Methods of test for flammability of textile fabrics when subjected to a small igniting flame applied to the face or bottom edge of vertically oriented specimens
- BS 5499. Fire safety signs, notices and graphic symbols
  Part 1. 1990. Specification for fire safety signs
  Part 2. 1986. Specification for self-luminous fire safety signs
  Part 3. 1990. Specification for internally-illuminated fire safety signs
- BS 5502. Part 23. 1990. Buildings and structures for agriculture. Code of practice for fire precautions
- BS 5588. Fire precautions in the design, construction and use of buildings
  Part 1. 1990. Code of practice for residential buildings
  [†] Part 2. 1985. Code of practice for shops
  [†] Part 3. 1983. Code of practice for office buildings
  [†] Part 4. 1978. Code of practice for smoke control in protected escape routes using pressurization
  Part 5. 1991. Code of practice for firefighting stairs and lifts
  Part 6. 1991. Code of practice for places of assembly
  [‡] Part 7. Code of practice for atrium buildings
  Part 8. 1988. Code of practice for means of escape for disabled people
  Part 9. 1989. Code of practice for ventilation and air conditioning ductwork (*revision of* Appendix A to CP 413)
  Part 10. 1991. Code of practice for shopping complexes
  [‡] Part 11. Code of practice for shops, offices, factories, warehouses and similar workplaces
- BS 5628. Part 3. 1985. Code of practice for use of masonry. Materials and components, design and workmanship
- BS 5803. Part 4. 1985. Thermal insulation for use in pitched roof spaces in dwellings. Methods for determining flammability and resistance to smouldering
- BS 5839. Fire detection and alarm systems for buildings
  Part 1. 1988. Code of practice for system design, installation and servicing

---

[†] In course of revision.
[‡] In course of preparation.

- Part 3. 1988. Specification for automatic release mechanisms for certain fire protection equipment
- BS 5852. Fire tests for furniture
    - Part 1. 1979. Methods of test for the ignitability by smokers' materials of upholstered composites for seating
    - Part 2. 1982. Methods of test for the ignitability of upholstered composites for seating by flaming sources
- BS 5908. 1990. Code of practice for fire precautions in the chemical and allied industries
- BS 5950. Part 8. 1990. Structural use of steelwork in building. Code of practice for fire resistant design
- BS 6203. 1982. Guide to fire characteristics and fire performance of expanded polystyrene (EPS) used in building applications
- BS 6266. 1992. Code of practice for fire protection for electronic data processing installations
- BS 6336. 1982. Guide to development and presentation of fire tests and their use in hazard assessment
- BS 6373. 1985. Glossary of terms relating to burning behaviour of textiles and textile products
- BS 6661. 1986. Guide for design, construction and maintenance of single-skin air supported structures
- BS 8110. Part 2. 1985. Structural use of concrete. Code of practice for special circumstances. (*part revision* of CP 110: Part 1)
- BS 8202. Coatings for fire protection of building elements.
    - Part 1. 1987. Code of practice for the selection and installation of sprayed mineral coatings
    - Part 2. 1992. Code of practice for the use of intumescent coating systems to metallic substrates for providing fire resistance.
- BS 8214. 1990. Code of practice for fire door assemblies with non-metallic leaves
- BS 8313. 1989. Code of practice for accommodation of building services in ducts. (*revision of* CP 413)
- PD 6503. Toxicity of combustion products
    - Part 1. 1990. General (ISO/TR 9122-1)
    - Part 2. 1988. Guide to the relevance of small-scale tests for measuring the toxicity of combustion products of materials and composites
- PD 6512. Use of elements of structural fire protection with particular reference to the recommendations given in BS 5588 'Fire precautions in the design and construction of buildings'
    - ¶Part 1. 1985. Guide to fire doors
    - Part 3. 1987. Guide to the fire performance of glass

International Organization for Standardization (ISO)[§]

- ISO/IEC Guide 52: 1990. Glossary of fire terms and definitions

---

[¶] Proposed for withdrawal
[§] Available from BSI Sales Department, Linford Wood, Milton Keynes, MK14 6LE

- [†] ISO 834. 1975. Fire resistance tests — Elements of building construction
- ISO 1182. 1990. Fire tests — Building materials — Non-combustibility test
- ISO 1716. 1973. Building materials — Determination of calorific potential
- ISO 3008. 1976. Fire resistance tests — Door and shutter assemblies
- ISO 3009. 1976. Fire resistance tests — Glazed elements
- ISO 3261. 1975. Fire tests — Vocabulary
- ISO 5657. 1986. Fire tests — Reaction to fire — Ignitability of building products
- ISO 5925. Part 1. 1981. Fire tests — Evaluation of performance of smoke control door assemblies — Ambient temperature test
- ISO 6309. 1987. Fire protection — Safety signs
- ISO 6790. 1986. Equipment for fire protection and fire fighting — Graphical symbols for fire protection plans — Specification
- ISO 6944. 1985. Fire resistance tests — Ventilation ducts
- ISO 8421. Fire protection — Vocabulary
    Part 1. 1987. General terms and phenomena of fire
    Part 2. 1987. Structural fire protection
    Part 3. 1989. Fire detection and alarm
    Part 4. 1990. Fire extinction equipment
    Part 5. 1988. Smoke control
    Part 6. 1987. Evacuation and means of escape
    Part 7. 1987. Explosion detection and suppression means
    Part 8. 1990. Terms specific to fire-fighting, rescue services and handling hazardous materials

Department for Education*

- Building Bulletin 7. Fire and the design of educational buildings (1988)

Department of the Environment and The Welsh Office*

- The Building Regulations 1985. Manual to the Building Regulations 1985. (1985) (*withdrawn*)
- The Building Regulations 1985. Mandatory rules for means of escape in case of fire (1985) (*withdrawn*)
- The Building Regulations 1991. Approved Document B. Fire safety (1991)
- [‖] Building Regulation and Fire Safety. Procedural guidance (June 1992) (*published in conjunction with the Home Office*)

Department of the Environment for Northern Ireland*

- [‡] The Building Regulations (Northern Ireland) 1990. Technical Booklet E. Fire [1993]

Department of Health*

- Firecode. Directory of fire documents (1987)

[†] In course of revision.
[*] Available from HMSO (except where indicated).
[‖] Available from the DOE Distribution Unit, PO Box 151, London, E15 2HF. Fax: 081-533 1618.
[‡] In course of preparation.

- Firecode. Policy and principles (1987)
- Firecode. Nucleus fire precautions recommendations (1989)
- Health Technical Memoranda (HTM):
  - † HTM 81. Firecode. Fire precautions in new hospitals (1987)
  - HTM 82. Firecode. Alarm and detection systems (1989)
  - † HTM 83. Fire safety in health care premises. General fire precautions (1982)
  - ‡ HTM 84. Firecode. Fire precautions in new old peoples homes
  - ‡ HTM 85. Firecode. Fire precautions in existing hospitals [1993]
  - † HTM 86. Firecode. Assessing fire risk in existing hospital wards (1987)
  - † HTM 87. Firecode. Textiles and furniture (1989)
  - HTM 88. Fire safety in health care premises. Guide to fire precautions in NHS housing in the community for mentally handicapped (or mentally ill) people (1986)
- Fire Practice Notes (FPN):
  - FPN 1. Firecode. Laundries (1987)
  - FPN 2. Firecode. Storage of flammable liquids (1987)
  - FPN 3. Firecode. Escape bed lifts (1987)
  - ‡ FPN 4. Firecode. Kitchens
  - FPN 5. Firecode. Commercial enterprises on hospital premises (1992)

Greater London Council ¶

- Code of practice. Means of escape in case of fire (1976)
- Play safe. A guide to standards in halls used for occasional stage presentations (1980)
- A guide to fire safety in exhibitions and similar presentations in hotel buildings (1972)
- Code of practice for pop concerts (1978)

Health and Safety Executive*

- Fairgrounds and amusement parks: A code of safe practice (1992). Booklet HS(G)81
- § Guide to general fire precautions in explosives factories and magazines (1990)

Home Office*

- Fire Precautions Act 1971:
  - Guide to fire precautions in premises used as hotels and boarding houses which require a fire certificate (1991)
  - Guide to fire precautions in existing places of work that require a fire certificate. Factories, Offices, Shops and Railway Premises (1989)
  - Code of practice for fire precautions in factories, offices, shops and railway premises not required to have a fire certificate (1989)
  - Fire Safety Management in Hotels and Boarding Houses (1991)
  - Fire Safety at Work (1989)

---

† In course of revision.
‡ In course of preparation.
¶ These publications are no longer available for purchase.
§ Available from the Health and Safety Executive Sales Point, Room 414, St Hugh's House, Stanley Precinct, Bootle, Merseyside, L20 3QY.
* Available from HMSO (except where indicated).

- [‡] Guide to the Fire Precautions (Places of Work) Regulations [1993]
- Draft guides to fire precautions in existing:
    - [‖] Hospitals (1982)
    - [‖] Residential care premises (1983)
- Guide to means of escape and related fire safety measures in certain existing houses in multiple occupation (1988) (*withdrawn*)
- Guide to fire precautions in existing places of entertainment and like premises (1990)
- Fire Prevention Guides
    No 1 Fire precautions in town centre redevelopment (1972)
    (*withdrawn, see* BS 5588: Part 10)
    No 2 Fire precautions in new single-storey spirit storages and associated buildings (1973)
    No 3 Fire fighting and fire precautions in automated warehouses (1974)
    No 4 Safe use and storage of liquified petroleum gas in residential premises (1976)
- Guide to Safety at Sports Grounds (1990)

London District Surveyors Association[∅]

- Fire Safety Guide No 1. Fire Safety in Section 20 Buildings (1990)
- Fire Safety Guide No 2. Fire Safety in Atrium Buildings (1989)
- Fire Safety Guide No 3. Phased Evacuation from Office Buildings (1990)
- Model rules of management for places of public entertainment (1989)
- Model technical regulations for places of public entertainment (1991)
- Rules of management for occasional licences (1989)

Loss Prevention Council

- [†] Code of practice for the construction of buildings (1992)
- Fire Prevention on Construction Sites (1992)
    (*published in conjunction with the Building Employers Confederation and the National Contractors' Group*)

Scottish Office*

- [§] Fire safety: new health buildings in Scotland (1987)
- Guide to means of escape and related fire safety measures in existing houses in multiple occupation in Scotland (1988)
- [†] Building Standards (Scotland) Regulations 1990. Technical Standards (1990)

[‡] In course of preparation.
[‖] Available from the Home Office Fire and Emergency Planning Department, Horseferry House, Dean Ryle Street, London, SW1P 2AW.
[∅] Available from the London District Surveyors Association Publications, PO Box 15, London, SW6 3TU.
[†] In course of revision.
[*] Available from HMSO (except where indicated).
[§] Available from the Scottish Office library.

Index

	Page
British Fire Prevention Committee (BFPC)	22,24,92,93
British Standards and Codes of Practice	
BS 459: Part 3. Fire-check flush doors	67
BS 476	(see Standard fire tests)
BS 2782	(see Standard fire tests)
BS 4422. Glossary of terms associated with fire	3,4,42,51,69,74
BS 5266. Emergency lighting	81
BS 5268. Fire resistance of timber structures	63,64
BS 5378. Safety signs and colours	80
BS 5395. Stairs, ladders and walkways	77
BS 5499. Fire safety signs, notices and graphic symbols	80,81
BS 5588. Fire precautions in the design, construction and use of buildings	63,67,74,76,77
Part 1. Residential buildings	75
Part 2. Shops	75,77
Part 3. Office buildings	75,77
Part 6. Places of assembly	75
Part 8. Means of escape for disabled people	74
Part 10. Shopping complexes	75
BS 5839. Fire detection and alarm systems for buildings	
Part 3. 1988 (Automatic release mechanisms)	69
BS 6336. Guide to development and presentation of fire tests	4
BS 6373. Glossary of terms relating to burning behaviour of textiles and textile products	4
BS 8214. Fire door assemblies with non-metallic leaves	68
CP 3: Chapter IV. Precautions against fire	
1948 (Houses)	20
Part 1: 1962 (Flats and maisonettes)	20
Part 1: 1971 (Flats and maisonettes)	20,67
Part 2: 1968 (Shops and departmental stores)	20,67
Part 3: 1968 (Office buildings)	20,67
CP 153: Part 4. Fire hazards associated with glazing	63
PD 6512: Part 1. Guide to fire doors	67
PD 6512: Part 3. Guide to the fire performance of glass	63
DD 64. Guidelines for the development and presentation of fire tests	4
Building Regulations	
England and Wales	18,28–30,103,104
Inner London	29,30

Northern Ireland	28,29
Scotland	18,28,29
Building Research Station	24,26
Charring rates (of timber)	63 and Table 11
Class 0	54
Comité Européen de Normalisation (CEN)	52
Construction of	
chimneys	9,10,11,13,29
external walls	8,9,10,11,12,29 and Figure 2
floors	11,12
party walls	8,11,12,13,29 and Figure 2
roofs	8,9,11,12,29,65
Couvre Feu	8
EEC Directives	
construction products	30
places of work	38
safety signs	80
Elements of construction	
beams	60
columns	60
floors	60,64,65
walls	60,64 and Table 10
Escape routes	
door fastenings	16,80
lighting	80,81
location	76
notices	16,80,81
number of	76,77
protection of	79
rate of movement	77
travel distances	79 and Table 13
width	77 and Table 12
European Community	30,38,52
Factories and warehouses	11,12,13,16,17,19,31,36,37,38,39
Fire	
Doors	11,50,65,67,68,69,80
Certificates	16,31,36,37,39,95,96,97,99
Compartmentation	17,29,57
Fighting facilities	9,11,12,13,17
Growth	42,45,53
Hazard	1,2
Ignition	42,43
Insurance	10,28,40
Losses	1 and Figure 1
Precautions	x,1,2,3
Prevention	3

Propagation indices	54,58,108,110 and Table 7
Protection	3
Protection (structural)	51
Resistance of elements of construction	24,29,49,50,57,60
Resisting materials	4,17,91
Spread (external)	57–59
(internal)	57
Statistics	1 and Tables 1–4
Tests	47–53,105–120

Fire Engine Establishments
Edinburgh	12
London	13

Fire Grading Committee   1,20,74,77
Fire Offices' Committee (FOC)   20,24,40
Fire testing
Association of Architects	21,22
Austria	22
British Fire Prevention Committee (BFPC)	22,24,92,93
Fire Offices' Committee (FOC)	24
Germany	22
Hartley (David)	21
Joint Fire Research Organization (JFRO)	26
Mahon (Viscount)	21
National Fire Brigades' Association (NFBA)	24
USA	22

Fires
1212 London	8,9
1666 London	1,9
1675 Northampton	11
1731 Blandford Forum	11
1743 Crediton	11
1824 Edinburgh	12
1861 Tooley Street, London	13
1881 Ring Theatre, Vienna	18,22
1887 Theatre Royal, Exeter	13
1897 Paris Charity Bazaar	16
1897 Cripplegate, London	16,22
1902 Queen Victoria Street, London	17
1903 Iroquois Theatre, Chicago	18,22
1911 Empire Palace Theatre, Edinburgh	18,19
1923 Grand Assembly Rooms, Leeds	18
1926 Drumcollagher, Co Limerick	18
1929 Glen Cinema, Paisley	18,19
1931 Coventry	18
1956 Eastwood Mills, Keighley	19
1960 Henderson's Department Store, Liverpool	19
1961 Bolton Top Storey Club	19
1969 Rose and Crown Hotel, Saffron Walden	19

1973 Summerland, Isle of Man	1
1979 Woolworth's, Manchester	1
1981 Stardust Disco, Dublin	1
1985 Bradford City football ground	1,20
1987 King's Cross	1,38

Flame retardants 48,56
Flashover 22,26,47
Guidance
1879 Metropolitan Board of Works (places of entertainment)	16
1934 Home Office (places of entertainment)	18
1935 Building Industries National Council	20
1945 Building Industries National Council	20
1946 Fire Grading Committee	1,20,29
1948 BSI (houses and flats of not more than two storeys)	20
1952 Fire Grading Committee	20,29,74
1962 BSI (high-rise flats and maisonettes)	20
1968 BSI (shops and departmental stores)	20,67
1968 BSI (office buildings)	20,67
1971 BSI (flats and maisonettes)	20,67
1972 Home Office (hotels and boarding houses)	36,75
1977 Home Office (factories)	36,75
1977 Home Office (offices, shops and railway premises)	36,75
1982 Home Office (hospitals)	75
1983 Home Office (residential care premises)	75
1985 DOE (Approved Document B)	29,54,63,67 and Table 10
1985 DOE (Mandatory Rules for Means of Escape)	29
1987 Department of Health	75
1988 Department of Education and Science (schools)	75
1989 Home Office (places of work requiring a certificate)	36
1989 Home Office (places of work not requiring a certificate)	38
1989 Home Office (for managers in places of work)	36
1990 Home Office (places of entertainment)	75
1990 Home Office (sports grounds)	20
1990 Scottish Office (Technical Standards)	29,75
1991 Home Office (hotels and boarding houses)	36
1991 Home Office (for managers in hotels and boarding houses)	36
1991 BSI (places of assembly)	75
1991 BSI (shopping complexes)	75
1991 DOE (Approved Document B)	29,75
1992 DOE (houses in multiple occupation)	40,75
1992 LPC (construction of buildings)	40
[1993] Home Office (places of work regulations)	38
[1993] DOE Northern Ireland (Technical Booklet E)	29

Hotels 13,17,19,31,36
Houses in multiple occupation 20,39,40
Ignition
source	43
temperature	42,43,44 and Table 6

International Fire Congress 1901	22
International Fire Prevention Congress 1903	24
International Organization for Standardization	
ISO/IEC Guide 52. Terminology	3
ISO 5657. Ignitability	112
ISO 5925–1. Smoke penetration	117
ISO 6309. Safety signs	80
ISO 6944. Ventilation ducts	114
ISO 8421. Vocabulary	3
Joint Fire Research Organization (JFRO)	26
$k \rho c$	44
Legislation	28–40
Limitation of	
cubic capacity	13,16
floor area	11,17
height	16
Linings	47,53,54,56
Materials	
concrete	18,50,62
glass	63
of limited combustibility	54
plastics	44,54,56
steel	12,18,50,62
timber/wood	44,50,56,62,63
Means of escape	74–81
considerations	75,76
ladders	11
requirements	13,16–20,75
routes	(see Escape routes)
Model Byelaws	
England	12,18
Scotland	18
National Fire Brigades' Association (NFBA)	24
Occupant capacity	77
Offices	31,36,37,38
Radiation	47,58
Regulations	
Building	(see Building Regulations)
Construction Products	30
Fire Certificates (Special Premises)	39,99–101 and Table 5
Fire Precautions (Non-certificated factory, office, shop and railway premises)	38
Fire Precautions (Places of Work)	31,38
Sub-surface Railway Stations	38
Reports	
1867 Select Committee on Fire Protection	13
1923 Royal Commission on Fire Brigades and Fire Prevention	17
1935 Building Industries National Council	20

1945 Building Industries National Council	20
1946 Fire Grading Committee. Post-War Building Studies No 20	1,20
1952 Fire Grading Committee. Post-War Building Studies No 29	20,74
Residential accommodation	13,16,18
Results of standard tests	53,59 and Tables 7 & 8
Roofs	
tests	26,48,59
thatch	8,9,59,65
venting	71,72
Royal Institute of British Architects (RIBA)	12,21,24
Sachs, Edwin	22,24
Shops	13,19,31,37,38
Smoke	
control	70,79
movement	69,70
Stages of a fire	41,42
Standard fire tests	
BS 476. Fire tests on building materials and structures	
(1932)	24,26,67
(1953)	26
Part 1	26
Part 2. Flammability	26
Part 3. Roofs (1958)	26,48,51,59,105,106,107
Part 3. Roofs (1975)	51,107
Part 4. Non-combustibility	26,53,107
Part 5. Ignitability	53,108
Part 6. Fire propagation	26,48,53,54,58,108,110
Part 7. Surface spread of flame	26,48,53,54,110,111
Part 8. Fire resistance	3,26
Part 10. Principles/application	51
Part 11. Heat emission	54,111
Part 13. Ignitability	112
Part 20. Fire resistance (general)	60,113,114
Part 21. Fire resistance (loadbearing elements)	60,113,114
Part 22. Fire resistance (non-loadbearing elements)	60,67,113,114
Part 23. Fire resistance (components)	60,113,114
Part 24. Fire resistance (ventilation ductwork)	114,116
Section 31.1. Smoke penetration	67,117
BS 2782. Methods of testing plastics	
Method 120A. Vicat softening temperature	118
Method 140D. Flammability	119
Method 140E. Flammability	119
Method 508A. Rate of burning	118
Statutes	
1189 London Assize	8
1246 London Assize	9
1667 London Rebuilding Act	9,10
1675 Northampton Rebuilding Act	11

1677 Southwark Rebuilding Act	11
1694 Warwick Rebuilding Act	11
1731 Blandford Forum Rebuilding Act	11
1731 Tiverton Rebuilding Act	11
1751 Disorderly Houses Act	12
1763 Wareham Rebuilding Act	11
1774 Fires Prevention (Metropolis) Act	11,12,21,22
1808 Chudleigh Rebuilding Act	11
1842 Liverpool Building Act	12
1843 Theatres Act	13,16
1843 Liverpool Fire Prevention Act	12
1844 Liverpool Fire Prevention Act	12
1844 Metropolitan Building Act	11,12
1847 Towns Improvement Clauses Act	12
1847 Town Police Clauses Act	12
1855 Metropolitan Building Act	12
1858 Local Government Act	12
1875 Public Health Act	12,17
1878 Metropolis Management and Building Acts (Amendment) Act	16
1882 Metropolitan Board of Works (Various Powers) Act	16
1888 Local Government Act	16
1890 Public Health Acts Amendment Act	12,13
1891 Factory and Workshop Act	16,17
1894 London Building Act	16,91
1895 Factory and Workshop Act	17
1901 Factory and Workshop Act	17,19
1905 London Building Acts (Amendment) Act	17
1907 Public Health Acts Amendment Act	17
1908 London County Council (General Powers) Act	18
1909 London County Council (General Powers) Act	18
1909 Cinematograph Act	19
1922 Celluloid and Cinematograph Film Act	Table 5
1928 Petroleum (Consolidation) Act	Table 5
1936 Public Health Act	18
1937 Factories Act	19
1939 London Building Acts (Amendment) Act	18,29,94,95
1944 Education Act	Table 5
1959 Factories Act	19
1961 Factories Act	19,31
1961 Housing Act	20
1961 Licensing Act	19
1961 Public Health Act	18
1963 Offices, Shops and Railway Premises Act	19,31
1963 London Government Act	Table 5
1964 Licensing Act	Table 5
1967 Private Places of Entertainment (Licensing) Act	Table 5
1968 Theatres Act	Table 5
1971 Fire Precautions Act	18,19,31,36–38,95–99 and Table 5

1974 Health and Safety at Work etc Act	31,39 and Table 5
1975 Safety of Sports Grounds Act	20 and Table 5
1980 Housing Act	20
1982 Local Government (Miscellaneous Provisions) Act	Table 5
1984 Building Act	101–103 and Table 5
1985 Cinemas Act	Table 5
1985 Housing Act	20,39,40 and Table 5
1987 Fire Safety and Safety of Places of Sport Act	20,31
1989 Local Government and Housing Act	39

Statutes (Northern Ireland)

1984 Fire Services Order	37 and Table 5

Statutes (Scotland)

1425 James I. Chapter 23	9
1674 Edinburgh Council	11
1698 Edinburgh Building Act	10
1850 Burgh Police Act	12
1892 Burgh Police Act	17
1897 Public Health Act	17
1900 Glasgow Building Regulations Act	17
1959 Building Act	18
1976 Licensing Act	Table 5
1980 Education Act	Table 5
1981 Education Act	Table 5
1982 Civic Government Act	Table 5
1987 Housing Act	40 and Table 5

Structural stability	60
Supervisory Schemes for fire test laboratories	52
Surface spread of flame	26,29,48,53,54,56 and Table 8
Terminology	3,4,24,42,74
Theatres and other places of assembly	11,13,16,17,18,19,22